DOWN TO EARTH MATHEMATICS

LOLA J. MAY • SHIRLEY M. FRYE

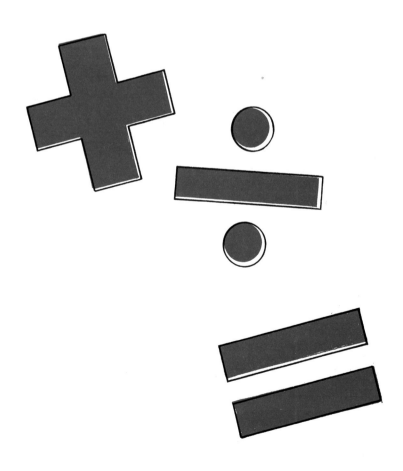

Copyright © 1995 by Didax Educational Resources, Inc.
All rights reserved.
Printed in the United States of America.

This book is published by
Didax Educational Resources, Inc., Rowley, Massachusetts.

Reproduction of any part of this book for use by other than the purchaser –
an entire school or school system – is strictly prohibited.

Except as noted above, no part of this publication may be reproduced,
stored in a retrieval system, or transmitted in any form or by any means –
electronic, mechanical, recording, or otherwise – without the prior
written permission of the publisher.

ISBN 1-885111-10-X

Designer/Illustrator: Collin Fry
Editor: Dee Corr

Down to Earth MATHEMATICS

INTRODUCTION

Simply stated, the aim of this book is to reinforce and supplement your teaching of the major mathematics objectives for grades K through five.

It is well known that there is no one way to teach mathematics, and teachers need to be willing to risk using different approaches to help their students understand the basic concepts of mathematics. To clarify concepts you usually need to use models and activities in addition to the material presented in the basic textbooks.

Down to Earth Mathematics is intended as a resource of activity-based learning experiences for you to use with students in a variety of ways. Many students do not completely understand a mathematical concept the first time it is presented, and, therefore, need different experiences using a new approach, beyond the initial exposure.

The STANDARDS documents of the National Council of Teachers of Mathematics are the vision for change, and this change means students should learn more than just computation. In the problem-solving approach students become aware that there is often more than one way to solve problems. Students also learn that some problems cannot be solved in a few minutes.

This book will give you guidance in helping students develop problem-solving approaches and will present activities you can use to assess the depth of their understanding of the mathematics concepts. We hope that the activities in this book will serve as a springboard for creating your own activities and sharing them with other teachers. In this way all students will become successful in understanding and enjoying mathematics.

Lola J. May

Shirley M. Frye

Table of CONTENTS

Chapter 1 — Place Value   1 – 19

- Numbers Zero Through Nine .... 2 – 4
- Tens and Ones 5, 6
- The Hundred Chart 7
- Hundreds, Tens, Ones .... 8 – 10
- Base Ten Models 11
- Expanded Notation 12
- Increase and Decrease Value of a Digit 13, 14
- Expanded Notation – Snap Game 15, 16
- Rounding Numbers to the Nearest Ten 17, 18
- Rounding Numbers to the Nearest Hundred 19

Chapter 2 — Addition/Subtraction   20 – 48

- Meaning of Addition .... 21 – 25
- Addition Facts to 15 – Concentration 25, 26
- Addition Facts – Solitary 27, 28
- Addition Facts – Make Twenty .... 28 – 30
- Addition Facts – More or Less 30, 31
- Addition – Mental Math 31, 32
- Addition – Adding Larger Numbers 33
- Palindrome 34
- Meaning of Subtraction 35, 36
- The Hill Method of Subtraction Facts 11 Through 18 .... 37, 38
- Triangular Flash Cards 39
- Subtracting Two-Digit Numbers 40, 41
- Subtraction Strategies 41, 42
- Problem Solving – Educated Guessing 43, 44
- Missing Digits – I 44
- Missing Digits – II 45
- Frame Arithmetic 46
- Other Methods of Addition and Subtraction 47, 48

Chapter 3 — Multiplication/Division   49 – 65

- Manipulative/Pictorial Models 50, 51
- Frame Arithmetic .... 51 – 53
- The Properties of Multiplication 53
- The Special Number 9 54
- Six Additional Facts 54
- Writing Complete Equations 55
- Triangular Flash Cards 56
- Multiply and Add 56
- High or Low Game 57
- Multiplying a Many-Digit Factor By a One-Digit Factor ..... 58 – 60
- Assessing the Meaning of Multiplication 61
- Casting Out Nines 62
- Division Readiness 63, 64
- Division Models 64, 65

Chapter 4 — Measurement — 66 – 77

- Non-Standard Units 67, 68
- Telling Time 68, 69
- Linear Measurement . . . . 69 – 74
- Perimeter . . . . 74 – 76
- Area 76, 77

Chapter 5 — Geometry — 78 – 88

- Three-Dimensional Shapes 79
- Two-Dimensional Shapes 80
- Right, Acute and Obtuse Angles 81, 82
- Angles in Geometric Shapes 83, 84
- Parallel and Intersecting Lines 84
- Intersecting, Parallel and Perpendicular Lines 85, 86
- Line Symmetry 87, 88

Chapter 6 — Fractions — 89 – 102

- Meaning of Fractions 90
- Naming Fractions . . . . 91 – 95
- Assessing the Understanding of Fractions 95, 96
- Equivalent Fractions . . . . 97 – 99
- Comparing and Ordering Fractions 100, 101
- Improper Fractions/Mixed Numbers 101, 102

Chapter 7 — Problem Solving — 103 – 113

- Patterns in Problem Solving 104
- Patterns With Objects 105, 106
- Organizing Data . . . . 106 – 108
- Word Problems 109, 110
- Number Puzzles 110, 111
- Small Group Problem Solving 111
- Games 112, 113

Chapter 8 — Number Sense — 114 – 128

- Odd and Even Numbers 115, 116
- Number Games . . . . 117 – 119
- Mental Computation . . . . 119 – 121
- Computational Estimation 121, 122
- Number Theory . . . . 122 – 128

Correlation to NCTM "Curriculum and Evaluation Standards for Teaching Mathematics" — Grades K – 4

Chapter 1 — Place Value

Numbers Zero Through Nine
Tens and Ones
The Hundred Chart
Hundreds, Tens, Ones
Base Ten Models

Expanded Notation
Increase and Decrease Value of a Digit
Expanded Notation – Snap Game
Rounding Numbers to the Nearest Ten
Rounding Numbers to the Nearest Hundred

Standards that correlate with these activities: 2, 3, 5, 6, 13
- Communication
- Reasoning
- Estimation
- Number Sense and Numeration
- Patterns and Relationships

Chapter 2 — Addition/Subtraction

Meaning of Addition
Addition Facts to 15 – Concentration
Addition Facts – Solitary
Addition Facts – Make Twenty
Addition Facts – More or Less
Addition – Mental Math
Addition – Adding Larger Numbers
Palindrome
Meaning of Subtraction
The Hill Method of Subtraction Facts 11 Through 18

Triangular Flash Cards
Subtracting Two-Digit Numbers
Subtraction Strategies
Problem Solving – Educated Guessing
Missing Digits – I
Missing Digits – II
Frame Arithmetic
Other Methods of Addition and Subtraction

Standards that correlate with these activities: 1, 2, 3, 4, 7, 8, 13
- Problem Solving
- Communication
- Reasoning
- Connections
- Concept of Whole Number Operations
- Whole Number Computation

Chapter 3 — Multiplication/Division

Manipulative/Pictorial Models
Frame Arithmetic
The Properties of Multiplication
The Special Number 9
Six Additional Facts
Writing Complete Equations
Triangular Flash Cards
Multiply and Add

High or Low Game
Multiplying a Many-Digit Factor By a One-Digit Factor
Assessing the Meaning of Multiplication
Casting Out Nines
Division Readiness
Division Models
Division Assessment

Standards that correlate with these activities: 2, 3, 4, 6, 7, 8
- Communication
- Reasoning
- Connections
- Concept of Whole Number Operations
- Number Sense & Numeration
- Whole Number Computation

Chapter 4 — Measurement

    Non-Standard Units   Perimeter
    Telling Time   Area
    Linear Measurement

Standards that correlate with these activities: 1, 2, 3, 4, 5, 6, 8, 9, 10, 11
- Communication
- Geometry & Spatial Sense
- Reasoning
- Number Sense & Numeration
- Estimation

Chapter 5 — Geometry

    Three-Dimensional Shapes   Right, Acute and Obtuse Angles
    Two-Dimensional Shapes   Angles in Geometric Shapes
    Parallel and Intersecting Lines   Faces, Edges and Vertices
    Intersecting, Parallel and Perpendicular Lines   Line Symmetry

Standards that correlate with these activities: 2, 3, 9, 10
- Communication
- Geometry & Spatial Sense
- Reasoning
- Measurement

Chapter 6 — Fractions

    Meaning of Fractions   Equivalent Fractions
    Naming Fractions   Comparing and Ordering Fractions
    Assessing the Understanding of Fractions   Improper Fractions/Mixed Numbers

Standards that correlate with these activities: 2, 3, 4, 12, 13
- Communication
- Patterns & Relationships
- Reasoning
- Fractions & Decimals
- Connections

Chapter 7 — Problem Solving

    Patterns in Problem Solving   Number Puzzles
    Patterns With Objects   Small Group Problem Solving
    Organizing Data   Games
    Word Problems

Standards that correlate with these activities: 1, 2, 3, 4, 11, 13
- Problem Solving
- Connections
- Communication
- Statistics & Probability
- Reasoning
- Patterns & Relationships

Chapter 8 — Number Sense

    Odd and Even Numbers   Computational Estimation
    Number Games   Number Theory
    Mental Computation

Standards that correlate with these activities: 1, 2, 3, 4, 6, 8, 12
- Problem Solving
- Connections
- Communication
- Number Sense
- Reasoning
- Whole Number Computation

PLACE VALUE
Chapter 1

BACKGROUND

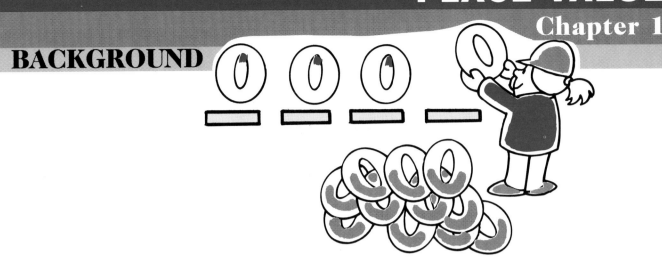

In order for students to understand the operations used in mathematics, they first must have a clear understanding of the formation of our number system. This system uses only ten symbols called digits – 0, 1, 2, 3, 4, 5, 6, 7, 8, 9. By using these ten symbols or digits and place value, we are able to write numbers of any quantity. The position of these digits determines the values that are associated with them. When we write 32, we mean 3 groups of ten and 2 ones. We can then say that 3 is in the *tens* place and 2 is in the *ones* or *units* place.

This place-value system leads to ease and convenience when performing mathematical computations. The development of a place-value system and the invention of the symbol for zero were crucial in the development of mathematics.

The following activities give students an opportunity to work with numbers in a concrete way to firmly establish the concepts of our place-value system.

1

PLACE VALUE

NUMBERS ZERO THROUGH NINE

Students learn to recognize and count numbers of objects through one-to-one correspondence. Since young students have used cardinal and ordinal numbers informally before attending school, it is important to build on their interests to introduce numeration concepts.

OBJECTIVE: Students create sets of objects for numbers 1 through 5 and match the sets to numeral cards.

ACTIVITY 1
Give each student 15 chips or markers, and ask the students to make sets of 1, 2, 3, 4 and 5 objects. Cards with numbers 1, 2, 3, 4 and 5 should be provided. Ask the students to match the numeral cards to each set they have made.

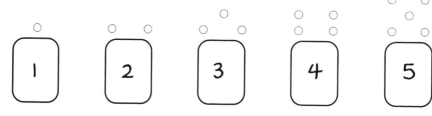

ACTIVITY 2
Ask the students to mix up the number cards then create a set of objects to match the numbers.

ACTIVITY 3
Using their fingers, students respond to the number you show and say. For example, show and say, "three." The students respond with three fingers.

ACTIVITY 4
Show a picture of 1, 2, 3, 4 or 5 objects quickly. Students respond by writing the number on a slate or wipe-off board. Asking the students to make the number in the air is another method.

PLACE VALUE

OBJECTIVE: Students create sets of objects for numbers 5 through 9 and match the sets to numeral cards.

ACTIVITY 1

For this activity each student will need 25 chips or markers of one color and 10 chips or markers of another color. Ask the students to use their 25 chips to make five sets of 5 chips each. Using the 5 configuration, students add more chips of the second color to create new sets up to 9 as illustrated.

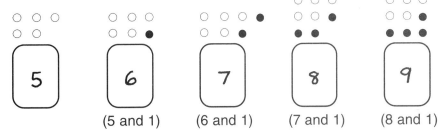

As the students add more chips to each set, they should say: "5 and 1 is 6, 6 and 1 is 7, 7 and 1 is 8, 8 and 1 is 9." As each new set is made, a numeral card can be placed under the set.

ACTIVITY 2

Students model a given number in many ways to build the concept of the number.

Example: Make in as many ways as possible. Using a folded sheet of paper as a work area, ask students to arrange six objects in as many different ways as possible.

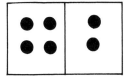 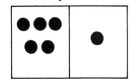 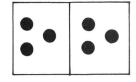 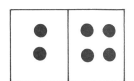 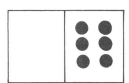

The students can say: "4 and 2 is another name for 6, 5 and 1 is 6, 3 and 3 is 6, 2 and 4 is 6, and 0 and 6 is 6." Repeat this activity for sets of 7, 8 and 9.

PLACE VALUE

ACTIVITY 3
In this activity, use illustrations of a standard set of objects such as domino patterns, instead of scattered objects on one side of a card. The other side of the card should have the numeral that matches the set.

The cards are placed in a mixed order and students are asked to place them in order one to nine, first using the side with objects illustrated and then repeat the activity using the numeral side.

  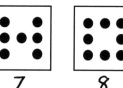

As an additional activity, the card with six objects on it can be placed on the table and the students are asked to show cards with one less and one more object on it. This activity can be repeated by showing any of the other numeral cards and asking the students to show cards with one less and one more.

OBJECTIVE: Students learn the names and sequence of the ordinal numbers through tenths.

ACTIVITY 1
Place stuffed animals, such as teddy bears, in a line of five bears. Starting at the left, ask the students to point to the first, second, third, fourth and fifth bear. Repeat this activity by starting at the right.

Repeat this activity by adding one more animal until there are ten animals in the line.

ACTIVITY 2
Ask ten students to line up in a row in front of the class. Then say, "Starting at the left, the third and fifth students please step forward." Repeat this activity starting at the right asking different ordinals to step forward.

PLACE VALUE

TENS AND ONES

Students first learn to count sets of ten and then sets of tens and ones as they learn the names of two-digit numbers. The concept of place value is being introduced and needs careful development with concrete materials.

OBJECTIVE: Students build sets of ten, count them and read the matching number.

ACTIVITY 1

Using objects, such as rods, ask students to form sets of ten. The sets of ten are counted in a manner similar to that used in counting ones. For example, students say, "one ten, two tens, three tens, etc." Cards with numbers 10, 20, 30, 40, etc. written on them can be matched to given sets of tens. Show a numeral card, such as, and ask the students to demonstrate the meaning by showing four sets of ten.

ACTIVITY 2

Given sets of objects of tens and ones, the students count and identify the number of tens and ones.

PLACE VALUE

Numeral cards can be matched to the sets when the students state the number of the set. The reverse of this activity would be to show the two-digit numeral card, and the students show the matching sets of tens and ones, such as

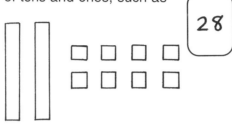

ACTIVITY 3
To assess the students' understanding of tens and ones, show two numbers such as 32 and 23. Ask the students how the numbers are different by creating sets of tens and ones.

ACTIVITY 4
Tell the students that there are "nicknames" for sets of tens and ones. One ten and three ones is called thirteen, two tens and five ones is called twenty-five. These names can be matched to a given set of numerals.

thirteen   twenty-five

PLACE VALUE

THE HUNDRED CHART

The hundred chart is a good visual aid to help students see patterns in numbers up to 100. The ability to see these mathematical patterns aids in the understanding of mathematics and facilitates computation.

OBJECTIVE: Students identify a number that is one less or one more than a given number, find number patterns and identify missing numbers.

ACTIVITY 1
Looking at the hundred chart, students identify which number is one less or one more than a given number.

Example: Which number is one less than 37? 56? 84? 70? etc.
What number is one more than 32? 45? 50? etc.

ACTIVITY 2
Use the hundred chart to search for the patterns in counting by ten.
Example: "3, 13, 23, 33, 43, ..."
Have students describe patterns they see in the chart.

ACTIVITY 3
Cover some of the numbers on the hundred chart and ask students to fill in the missing numbers.

1	2	3	4	5	6	7	8	9	10
11	12	13	14	15	16	17	18	19	20
21	22	23	24	25	26	27	28	29	30
31	32	33	34	35	36	37	38	39	40
41	42	43	44	45	46	47	48	49	50
51	52	53	54	55	56	57	58	59	60
61	62	63	64	65	66	67	68	69	70
71	72	73	74	75	76	77	78	79	80
81	82	83	84	85	86	87	88	89	90
91	92	93	94	95	96	97	98	99	100

49	
	60
	70

35		37
	47	48

PLACE VALUE

HUNDREDS, TENS, ONES

The pattern of 10 tens creates a model of a hundred. The concept of place value of each digit in a three-digit number requires development with concrete materials. The concept of expanded notation is introduced to show that the total value of each digit added together creates the total value of the whole number.

OBJECTIVE: Students represent one hundred as 10 tens.

ACTIVITY

A model of the 1 hundred as 10 tens or 100 ones can be shown in several ways:
- Ten bundles of ten objects to create a bundle of one hundred.
- Ten students hold up their ten fingers as the other children count: 1 ten, 2 tens ... 9 tens, 10 tens or a hundred fingers.
- Students use calculators to show 10 + 10 + 10 + 10 + 10 + 10 + 10 + 10 + 10 + 10 to show 100 on the display.
- Using the edges from computer paper, cut the strips into shorter strips of 10 holes each.
- Demonstrate that 10 strips of 10 holes represent a hundred holes.

OBJECTIVE: Students read and demonstrate three-digit numbers.

ACTIVITY 1

Show students numbers in a place-value chart, for example the number 345. Ask the students: "What is the total value of each digit in the number?"

hundreds	tens	ones
3	4	5

TOTAL VALUE?

Some responses can be: "3 is in the hundreds place, it means 300." "4 is in the tens place, it means 40." "5 is in the ones place, it means 5 ones." The number is read, "Three hundred forty-five."

PLACE VALUE

In continuing this activity, some of the numbers should show zero in the tens and ones place. In the number 407, students will see that the zero in 407 means no tens.

It is necessary to emphasize that zero is holding a place. Without zero, the number would be 47.

hundreds	tens	ones
4	0	7

ACTIVITY 2

Using 30 large cards, write the number 100 on ten cards, write 10 on another ten cards, and write 1 on ten cards. Distribute the cards to the students and then write a three-digit number such as 254 on the chalkboard. Holding the number cards, ask students to form columns in front of the digit to show the value of each digit.

Ask: "How much is the two worth?" (200) "The 5?" (50) "The 4?" (4) Continue this activity using three-digit numbers with zero as one of the digits.

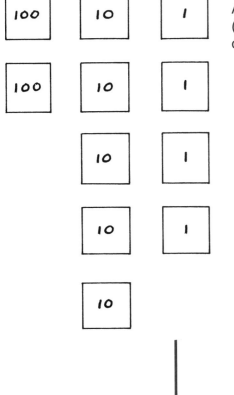

PLACE VALUE

ACTIVITY 3
Use the same cards and the same activities as explained in Activity 2 to decrease or increase the value of a number. Write the number 362. Ask the students to show the value of the number with the cards.

Now, say: "Decrease the number by 30 or 3 tens." Three students holding the 10 cards sit down and another student comes up and changes the number to 332. Then say, "Increase the number by 200." Two students holding 100 cards come up and join the column holding 100 cards. Another student changes the number on the board to 532.

ACTIVITY 4
In this activity, display models for hundreds, tens and ones in random order and ask students to sort them according to value in order to find the number represented.

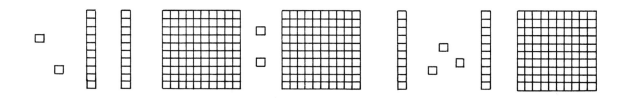

As an additional activity, have pairs of students do this activity. One shows the model and the other student names the number. For example, this group of place-value materials represents the number 347.

10

PLACE VALUE

BASE TEN MODELS

In primary grades most students have been involved with models that helped them understand our base ten system of numeration. Teachers usually used two main models – the bundling model and the area model. These models are also useful for reinforcing and reviewing base ten concepts with third and fourth grade students.

The bundling model is the least abstract of the two models and can be demonstrated using straws, sticks, rods or craft sticks. Using bundles of tens and hundreds, students learn that each new unit group or bundle is ten times as large as the previous unit. They can verify by counting. Each bundle can be opened to reveal the 10 units. Each ten bundle has 10 ones. Each hundred bundle has 10 tens.

The area model is more abstract than the bundling model because it is a geometric representation of number. Using sets of Base Ten Blocks, students discover that the ten rod is ten times as long as the ones cube; the hundred flat is ten times the size of the ten rod; the thousand cube is ten times the size of the hundred flat. The units cannot be separated but can be matched.

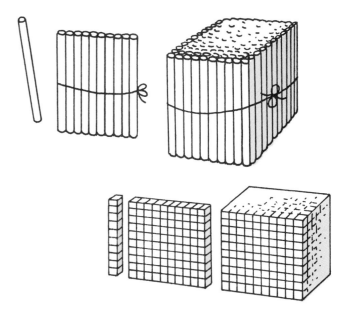

The size of the large thousand cube can be created by putting together 10 of the hundred squares or flats, illustrating that the pattern continues – 10 ones become a ten, 10 tens become a hundred, and 10 hundreds become a thousand.

The Base Ten Blocks can be used to help students understand total value in computation of whole numbers. The blocks can be placed in scrambled order and students can be asked to give the total value of the number represented. In order to do this, the student must combine the representations of thousands, hundreds, tens and ones. For some students, the first step will be placing blocks together.

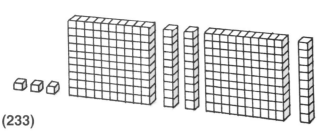

(233)

Two hundred thirty-three

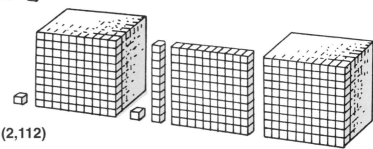

(2,112)

Two thousand one hundred twelve

11

PLACE VALUE

EXPANDED NOTATION

During mathematics instruction, students should indicate the place value of digits. In these activities you can assess if students really have a firm grasp of the important numeration concepts by asking the students to identify place and total value and to change the values.

Expanded notation in the vertical form should be presented before the horizontal form. The following activities should be demonstrated and then additional independent activities can be done by the students until you are confident that students have an understanding of expanded notation.

OBJECTIVE: Students identify place and total value of digits and express numbers in expanded notation.

Ask students to identify place value above each digit and total value below.

Place Value

(1,000)	(100)	(10)	(1)
8,	7	6	3

Total Value

(8,000)   (700)   (60)   (3)

Show students how to write these numbers in vertical expanded notation:

$$\begin{array}{r} 8000 \\ 700 \\ +60 \\ 3 \\ \hline \end{array}$$

Later students can learn to write expanded notation in the horizontal form:

$$8,763 = 8000 + 700 + 60 + 3$$

If horizontal notation is introduced before the vertical expanded notation, students may incorrectly give the answer as the sum of the digits. (8 + 7 + 6 + 3)

Vary the activity by having students call out four total values. Others record the number.

PLACE VALUE

INCREASE AND DECREASE VALUE OF A DIGIT

After using concrete models of place value, students can become involved in "Live Action Place Value." The entire class participates and the students change roles. For this activity, you will need a paper plate for each student in the class. On ten of the plates, write the numerals 0 through 9 in blue, on another ten plates write the numerals 0 through 9 in red and on ten other plates the numerals 0 through 9 in green. The number of plates should be adjusted so that each student in the classroom has one plate.

OBJECTIVE: Students increase and decrease the value of digits in a number.

ACTIVITY

Distribute the prepared plates to the students. Explain that you are going to call out digits. If they are holding one of the digits called, they are to come to the front of the room.

Call out colors and digits such as: "Red seven, blue five, and green nine. Please come up to the front of the class." Then ask the students to arrange themselves to form the greatest possible number and show their plates to the class.

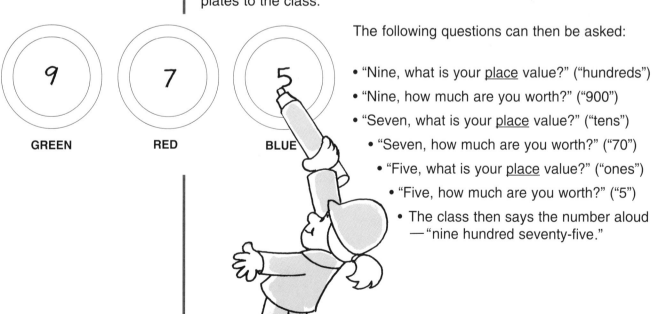

GREEN   RED   BLUE

The following questions can then be asked:

- "Nine, what is your <u>place</u> value?" ("hundreds")
- "Nine, how much are you worth?" ("900")
- "Seven, what is your <u>place</u> value?" ("tens")
- "Seven, how much are you worth?" ("70")
- "Five, what is your <u>place</u> value?" ("ones")
- "Five, how much are you worth?" ("5")
- The class then says the number aloud — "nine hundred seventy-five."

PLACE VALUE

Next, ask the students to form the least number.

| BLUE | RED | GREEN |

Once again, each student is asked to state the place value or total value of his or her digit and the class says the number aloud.

With students in the same order, 579, the number can be increased or made greater by 300. Ask the students: "Which student should sit down and which number should take his or her place to increase the number by 300?" Blue five sits down and blue eight comes up to make the number 879.

BLUE RED GREEN

Then ask the students to decrease the number 879 by 40. Red seven sits down and red three comes to form the number 839.

 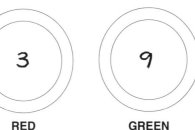

BLUE RED GREEN

This activity can be continued with students taking turns calling out the values to be increased or decreased or numbers to be created. Students can also practice writing the numbers in words.

> Remind students that "and" is not used in reading whole numbers. "And" is only used for reading the decimal point – "12 and 3 tenths."

EXTENSION

Ask students to make these numbers and verify their choices:
- even or odd numbers
- numbers with like digits
- least 3-digit number
- largest 3-digit number

Then ask: "How many numbers can be made with three different digits?"

PLACE VALUE

EXPANDED NOTATION — SNAP GAME

Numeration can be made more interesting and enjoyable for students through game activities such as "Snap." This activity will show you whether students have a good grasp of place value and total value of digits in numbers. One of the positive features of these activities is that all students have a chance to win.

OBJECTIVE: Students make, read and compare numbers in an activity called "Snap."

"Snap" can be played by the whole class or groups.

ACTIVITY — SNAP

Playing cards (four players use one deck) or number cards and an 8½" x 11" sheet of paper for each player are needed. Each student will need the following playing cards:

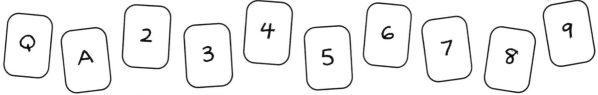

The Queen represents the number zero, and the Ace the number one. If playing cards are not available, number cards can be used. Each student also will need an 8½" x 11" sheet of paper or card stock to make his or her own playing board.

Ask the students to place the sheet of paper or card stock in a horizontal position and place a playing card in a vertical position in the upper left-hand corner of the paper and trace around the card. The card is then moved to the right and traced around until there are four rectangles. On the bottom right-hand side of the paper, students trace around the card and write "Discard" in the rectangle. Trace around the card in the middle of the playing board and write "Card Pile" in the rectangle.

PLACE VALUE

DIRECTIONS FOR PLAY

- Each student shuffles his or her cards and places them facedown on the Card Pile space.
- The four rectangles at the top of the playing board represent the thousands, hundreds, tens and ones places.
- In the first game students are to play for the greatest possible number.
- When you say, "Snap!" each student turns over the top card of his or her pile and places the card with the number up in one of the five rectangles.
- Remind the students that one card is placed in each rectangle and that once a card is put in place, it cannot be moved.
- Tell the students that you will say "Snap!" five times. For example, if the card turned up is 7, it could be placed in the hundred place. "Snap!" again, the card is 5. Place the five card in the tens place. "Snap!" again, the card is the Queen.
- Place the Queen card (zero) in the discard box. "Snap!" again, the card is 3. Place the 3 card in the ones place. Last "Snap!" — the number is 9. Place the 9 card in the thousands place.
- Ask the students to read their numbers. Students should read their numbers as: "nine thousand, seven hundred fifty-three." The student with the greatest number is the winner.

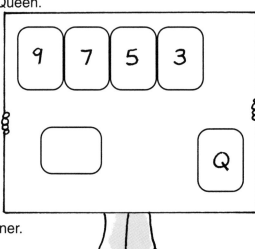

VARIATIONS

- Students try to form the least number. Ask students to share their hints on playing, such as: If the Queen is turned up, it can be placed in the thousands place. If nine is turned up, it should be placed in the discard box.
- Students try to create the greatest number that is even or odd, or the least number that is even or odd. For the even number, the ones place has to be 0, 2, 4, 6 or 8. For the odd number, the number 1, 3, 5, 7 or 9 must be in the ones place.
- Students can practice reading decimals by drawing a decimal point on the playing board.
- The activity can be adapted by having three places at the top of the playing board instead of four places.

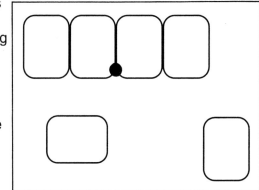

16

PLACE VALUE

ROUNDING NUMBERS TO THE NEAREST TEN

Estimation is used daily in our lives, and we actually do mental estimation during many of our regular activities. The STANDARDS of the National Council of Teachers of Mathematics stress the importance of estimation in learning mathematics.

One of the prerequisites of estimation with whole numbers is rounding numbers to multiples of ten and hundreds. A number line or a hundred chart can be used as a model for visualizing the relative location and size of numbers.

OBJECTIVE: Students will round numbers to the nearest ten.

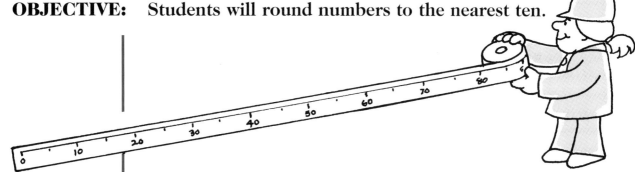

ACTIVITY

- Write the following on the chalkboard and ask the students to copy the numbers on their papers and then write the nearest ten before and after each number.

```
____ 72 ____
____ 18 ____
____ 45 ____
____ 53 ____
```

- After the tens are filled in by the students, draw frames around the ones digit in each number on your chalkboard example.

```
70  7|2|  80
10  1|8|  20
```

17

PLACE VALUE

Ask students: "Is the two in 72 five or more?" If the students respond, "No," then the number rounds down to 70. "Is the eight in 18, five or more?" If the students respond, "Yes," then the number rounds up to 20.

Discuss the location of the number on the number line and the relation to the nearest tens. After discussion, ask students to provide rules for rounding. Present the convention of "rounding up" when the digit is 5 or more.

Practice with a variety of 2- and 3-digit numbers. Use the number line until students can visualize without it.

6[4] ROUND _down_ TO _60_
3[7] ROUND _____ TO _____
7[5] ROUND _____ TO _____
32[9] ROUND _____ TO _____
48[1] ROUND _____ TO _____

- Write this example on the chalkboard with the ones in frames. Ask the students to copy the numbers and write "up" or "down" and then the rounded number.

- Using large numbers, students need to learn to round the number to the nearest ten. Ask the students to round 468,587 to the nearest ten. They should cover the number through the hundreds place.

[   ] 87

- Ask the students to round 87 to the nearest ten. The students' response should be "90." Then the number is uncovered and the students write the rounded number ending in 90.

This example shows the students that no matter what the size of the number, when you round to the nearest ten, you look at the tens and ones digits. Ask students what will always be true when rounding to the nearest ten. (Zero will always be in the ones place.) Use the number line to illustrate all the possible numbers that could be rounded to 60. (55, 56, 57, 58, 59, 61, 62, 63, 64)

You can assess students' understanding of rounding to the nearest ten by using a reverse activity. Ask each student to write three numbers that when rounded to the nearest ten is 80. The response would be between 75 and 84.

PLACE VALUE

ROUNDING NUMBERS TO THE NEAREST HUNDRED

Stress that the same steps used for rounding numbers to the nearest ten are used in rounding to the nearest hundred. Have students discuss their ways of rounding and when rounding would be used. Bring articles from newspapers and put a ring around the numbers. Decide which are exact and which are rounded numbers.

OBJECTIVE: Students round numbers to the nearest hundred.

Examples:

___ 468 ___
___ 617 ___
___ 348 ___

ACTIVITY

The activities in the preceding Rounding Numbers to the Nearest Ten section of this chapter can be adapted for rounding to the nearest hundred.
Is 468 closer to 400 or 500 on the number line?

```
|----------|----------|----------|
400        450      ↑ 500
                   468
```

Ask your students to write hundreds on each side of the numbers. Frame the digit in the tens place. Round up or down.

ROUND UP OR DOWN

400 4[6]8 500
600 6[1]7 700
800 8[4]8 900

Ask the students, "Is six in 468 five or more tens?" Students should answer, "Yes, round up to 500." "Is one ten in 617 five or more tens?" "No, round down to 600." "Is four tens in 848, five or more tens?" "No, round down to 800." Continue with other activities as found in the Rounding to Ten section.

GROUP ACTIVITIES

Students can work together in small groups to learn more about numeration systems using the following activities:

- Study Roman Numerals or Egyptian numbers and ask students to show how to write numbers with these different symbols and values. Show how these systems are alike and how they are different from our base ten system.

- Ask the students to make up a numeration system of their own. First, the students must decide on a base and then create the symbols that will represent the numbers.

- Present the students with the task of using only the keys [0] [1] [+] on the calculator to find the fewest number of inputs with the [+] necessary to make 333 on the display. Ask the students to explain how the task was accomplished. Continue with other challenges — 507, 628, 471.

19

ADDITION/SUBTRACTION
Chapter 2

BACKGROUND

The operations of addition and subtraction can both be modeled and practiced with part-part-whole. With manipulatives, addition is readily learned as combining part-part to get the whole while subtraction is developed as finding a part where the whole and one part are known. The relationship of these two operations can be demonstrated in the family of the three related numbers.

These fact families, or basic addition and subtraction facts, are necessary tools. Understanding the meaning of the operations precedes the memorization of the facts. Mental mathematics requires a knowledge of the facts that are used in individual ways by the students.

Calculators and computers are valuable learning and problem-solving tools, but they do not replace appropriate mental mathematics and the learning of the basic facts.

Students memorize the facts at different rates and with different strategies. There are some rules of mathematics that make the learning of all the basic facts easier such as the *order property* (commutative) of addition which allows the order of addends to be changed without changing the sum.

Another rule concerns zero which is called the *identity element*. When zero is one of the addends in an addition problem, the sum is the same as or identical to the other addend. Since 19 of the 100 addition facts have zero as an addend, this is a useful rule.

Understanding the place values and total values of digits in a number is also very important in addition. This knowledge is used when trading ones to tens and tens to hundreds. Concrete models that were presented in Chapter 1 should always be available for students in order to reinforce meanings.

The operation of subtraction is the inverse of addition and requires careful development. Subtracting large whole numbers has two prerequisites. The first is that the students have the basic facts of subtraction under control and be able to use the facts without any type of counting. Secondly, the student must understand the meaning of place value and total value of the digits in a number in order to trade.

The meaning of subtraction is taught using several models. The most common model is called "take away" in which some of the objects are taken from a whole set to find the other part. "Add on" is a model in which objects are added on to a given part to make the total set. The family of three numbers that shows the relationship of the facts helps students learn the addition and subtraction facts.

Assessment activities allow students to demonstrate their understanding of subtraction concepts by having them apply their knowledge of those concepts. This is an important strategy in the teaching of mathematics. If, for example, your students really understand subtraction, they will have no problem with the frame arithmetic activity. If they do have difficulty with this activity, more work with the subtraction process will be necessary.

ADDITION

MEANING OF ADDITION

The meaning of addition is taught most effectively through activities which combine two sets of objects to create a larger set. The basic addition facts are learned by recording the action performed with the materials used to model the facts. Students need to write the whole fact after each activity is completed. The following activities can be used to help your students learn the basic addition facts or to assess their understanding of addition concepts. The activities begin with the simplest manipulation of counters to more advanced activities.

OBJECTIVE: Students join two sets of objects to find the total number of objects.

ACTIVITY 1

In this activity, each student needs a part-part-whole frame and seven small counters. Tell a story such as: "The balloon man has four red balloons and three blue balloons. How many balloons altogether?"

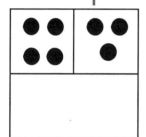 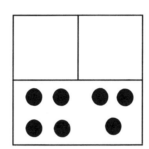

Students place four chips in one section of the frame and three chips in the other section. The chips are pulled down and are counted one by one.

ACTIVITY 2

In this activity, students place ten counters on the bottom section of the frame.

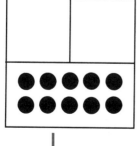 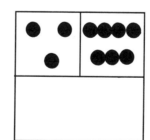 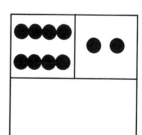

ADDITION

Each student separates the whole into the part-part sections and then records the different ways after each move. Students compare and discuss their charts.

10	
Part	Part
7	3
8	2

ACTIVITY 3

Rather than use counters only, in this activity students use numeral cards zero through nine and chips or counters. Tell a story such as: "You have six pennies and two pennies. How many pennies do you have altogether?"

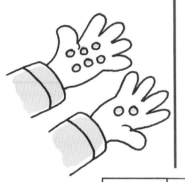

Students place the numeral 6 card in one section of the frame and two counters in the other section. The card and two counters are moved down to the bottom section. Students count 6, 7, 8 and then replace the combined parts with the numeral 8 card.

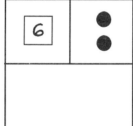 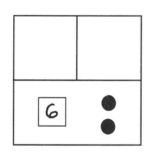 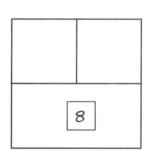

Repeat this activity with other numbers. Ask the students to use the numeral card for the larger number in each of the activities. As in the previous activity, students should write each addition fact.

OBJECTIVE: Students are given a sum and find all the facts that have that sum and record their equations.

ACTIVITY 1

Students use objects such as chips or counters for the given sum. For example, using six objects students form two sets in as many different ways as possible and record the facts as the sets are made. The activity can be repeated for the numbers 5, 7, 8, 9 and 10.

ADDITION

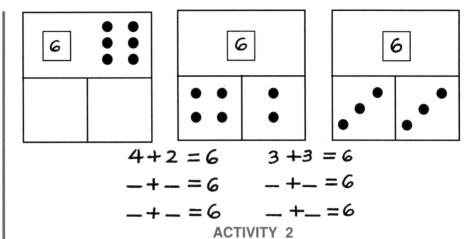

ACTIVITY 2
Using pictures of dominoes, students write the whole fact by using the dots on each side of the domino. Then students turn each domino around for another fact with the same sum, demonstrating the order property. The vertical form of facts can also be reinforced by changing the position of the dominoes.

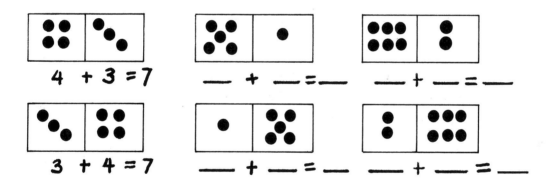

OBJECTIVE: Students write facts that are one more or one less than the given fact.

ACTIVITY
Using counters or chips, students create two sets of four, and write the fact 4 + 4 = 8. Then ask them to add one counter to one of the sets of four and write the new fact 4 + 5 = 9. Two sets of four are formed again, however, this time, students take out one counter from one of the sets of four, and write 4 + 3 = 7. This activity should be repeated with other doubles – 3 and 3, 5 and 5, 6 and 6, 7 and 7, 8 and 8.

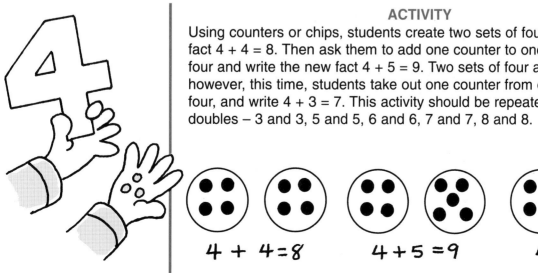

ADDITION

OBJECTIVE: Students learn that when one of the addends is zero, the sum is the same as the other addend.

ACTIVITY
Using calculators, ask the students to find the sum of 54 + 0, 0 + 86, 0 + 73, 18 + 0, 0 + 300, etc. After doing these exercises ask the students to explain what they have learned about adding zero to a number. Check their rule by using any number and zero as addends.

OBJECTIVE: Students practice sums of numbers through the sum of ten.

ACTIVITY
Each student needs 21 small index cards as illustrated below. The numbers written on the cards should fill the whole card.

ADDITION

The cards should be shuffled and placed faceup in two rows of five. Place the remaining cards on top of each of these cards. The twenty-first card can be placed on top of any pile.

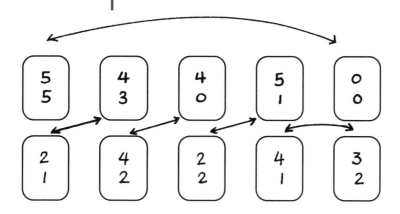

Ask students to pick up two cards, and only two cards, whose sum adds to 10. For example cards (5, 5) and (0, 0) add to 10. These cards are then removed. Cards (4, 3) and (2, 1) add to 10. Cards (4, 2) and (4, 0) add to 10.

The check card is the twenty-first card. This card should add to five. If the last card does not add to five, then an error has been made.

Suggest that the students crisscross the pairs of cards in the stack as they remove them, so that if an error occurs, they can find the pairs more easily.

ADDITION FACTS TO 15 — CONCENTRATION

The game of Concentration encourages students to practice the addition facts in an enjoyable activity while also fostering memory skills. The game is intended for two players. Playing cards or number cards are needed for this activity. The cards needed are 3 eights, 3 sevens, 3 nines, 3 sixes, 2 tens and 2 fives.

OBJECTIVE: Students learn the addition facts for a given sum.

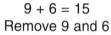
ADDITION

9 + 6 = 15
Remove 9 and 6

8 and 6 are not 15;
turn cards facedown

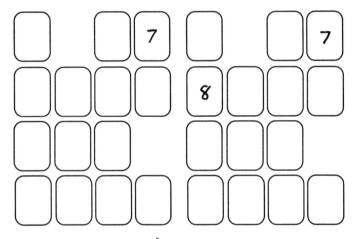

ACTIVITY
In this activity, one player shuffles the 16 cards, and lays out the cards facedown in four rows of four.

The 15 Game
The player that laid out the cards goes first by turning over two cards. If the two cards have numbers whose sum is 15, the player removes the cards and keeps them. The player continues to turn up two cards until the two cards do not add up to 15. If the cards do not add to 15, the player turns the cards facedown in the original places in the array.

When the player turns over two cards that do not have a sum of 15, it is the other player's turn. The second player turns over only one card. For example, if the card turned over is 7, the player now tries to remember where the 8 was in the array. If the player remembers and turns over the 8, he/she removes the 7 and 8 cards and gets another turn until the cards do not add to 15. If a player picks up and keeps two cards that do not add to 15, the other player can say "Challenge." If the challenge is correct, the player loses his or her turn and returns the cards to the playing area.

Play continues until all the cards are removed. The player with the most cards at the end of play wins.

For practice with sums of 11, 12, 13 and 14 use the following number of cards:

Sum of 14:   3 eights, 3 sixes, 3 nines, 3 fives, 2 tens and 2 fours
Sum of 13:   3 eights, 3 fives, 3 sevens, 3 sixes, 2 nines and 2 fours
Sum of 12:   3 eights, 3 fours, 3 sevens, 3 fives, 2 nines and 2 threes
Sum of 11:   3 eights, 2 threes, 3 sevens, 3 fours, 2 nines and 2 twos

ADDITION

ADDITION FACTS – SOLITARY

This activity is similar to Concentration except that the game is played by one player, the cards are placed faceup and the student finds all pairs of cards that have the sum of 11. The student will need a deck of cards with all face cards and Jokers removed. The Ace card equals one.

OBJECTIVE: Students learn the addition facts for a given sum.

ACTIVITY

The student should shuffle the cards and lay nine of them faceup in three rows of three. The player then picks up two cards, and only two cards, that have a sum of 11. In this example, the player could remove 9 and 2, 10 and Ace, 8 and 3 and 6 and 5. The only card left would be 4.

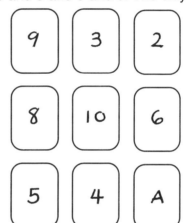

The student uses the deck of cards to replace the cards and make a complete array, leaving the remaining 4 in the same position. The student continues to remove two cards that add to 11, and continues making arrays of 9. The game ends when all the cards are removed. If there are no cards that add to 11, the student can add a fourth row of cards to increase the chances of making the sum of 11. Students should repeat the Solitary game until they can find the sums very quickly.

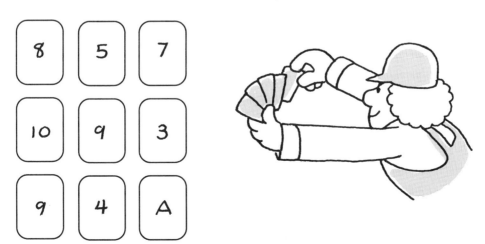

ADDITION

VARIATIONS

Sum of 12 –
Add four Jacks to the deck of cards. The Jack equals 11 and can match with the Ace which equals one.

Sum of 13 –
Add four Queens and keep the Jacks in the deck of cards. The Queen equals 12 and matches with the Ace to make 13. The Jack then matches with 2.

Sum of 14 –
Add four Kings and keep the Jacks and Queens in the deck of cards. The King equals 13 and matches with the Ace, the Queen equals 12 and matches with 2, Jack equals 11 and matches with 3.

ADDITION FACTS – MAKE TWENTY

Make Twenty is a solitary game. Prepare 45 index cards (3" x 5") for each student who will be participating in this activity. With cards in a vertical position, write two large numbers on each card with a marking pen.

The following numbers are written on each card. The arrow ➡ means to continue in a sequence. For example: (9,9), (9,8), (9,7), (9,6), (9,5), (9,4), (9,3), (9,2) and (9,1). There are no zeros.

After the cards are prepared, cut off the left top corner of each card to keep the deck in order.

OBJECTIVE: Students practice addition facts and increase the skill of scanning.

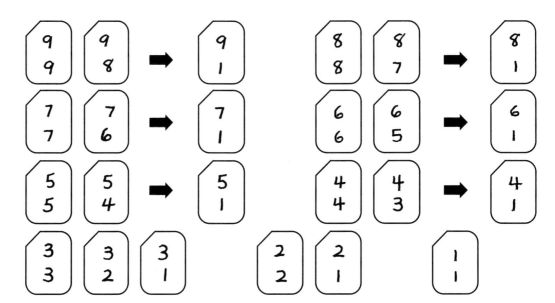

ADDITION

ACTIVITY

The student places the cards with numbers faceup creating two rows of five cards each. The student continues to place cards on top of the cards in each row. After all the cards are in place, each pile in the top row will have five cards and the bottom row piles will have four cards each.

The student picks up two cards whose numbers have the sum of 20. For example, (9,5) and (3,3) make 20. (8,7) and (4,1) make 20. (6,4) and (7,3) make 20. (8,4) and (5,3) make 20.

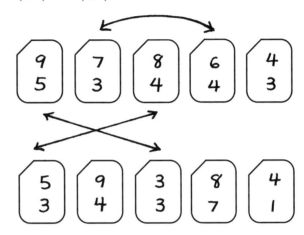

Students need to see that the matching cards can be in the same or different row. When cards are removed, it is best to keep the pairs of cards together by crisscrossing the cards that make 20 in their stacks, making it easier to check the sums if an error is made.

If a hole or empty space occurs, the student moves a card from one of the other piles, and tries to keep two rows of five cards going until only a few cards are left.

If students become stuck and can't find two cards whose sum is 20, he/she can "cheat" and move cards around in one pile until the play can resume.

The forty-fifth card is the last card in play and is called the Check Card. The Check Card will always have a sum of ten if no errors have been made. If the last card adds to a sum other than ten, an error has been made and the player should check back in the stack to find the error.

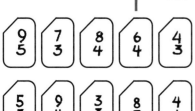

29

ADDITION

VARIATIONS

Students who have difficulty with the addition facts can work in pairs and alternate turns to remove two cards whose sum is 20.

The game can be made easier if the student makes two rows of six or seven cards. The more cards exposed, the easier the task.

Many math students often want more of a challenge. To make the game more challenging, remove (9,9), (9,8), (9,7), (9,6), (9,5), (9,4), (9,3), (9,2), (9,1). The objective of the game now is to find two cards whose sum is 18. This is more challenging than finding the sum of 20.

ADDITION FACTS — MORE OR LESS

The More or Less game is a two-player game using one set of the 45 cards that were created for the Make Twenty game.

OBJECTIVE: Students find sums and order the numbers.

ACTIVITY

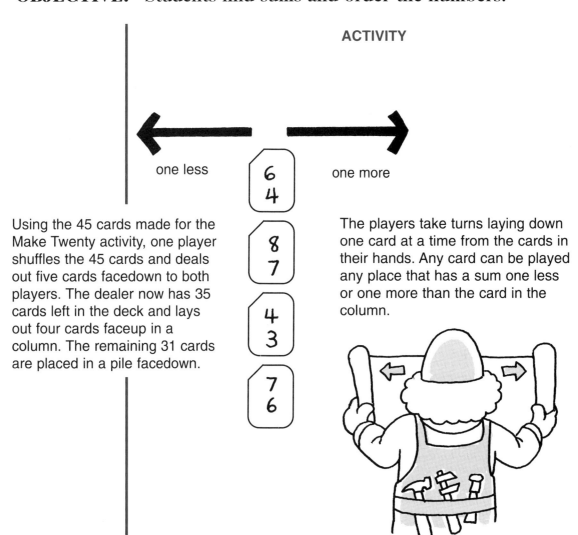

one less   one more

Using the 45 cards made for the Make Twenty activity, one player shuffles the 45 cards and deals out five cards facedown to both players. The dealer now has 35 cards left in the deck and lays out four cards faceup in a column. The remaining 31 cards are placed in a pile facedown.

The players take turns laying down one card at a time from the cards in their hands. Any card can be played any place that has a sum one less or one more than the card in the column.

ADDITION

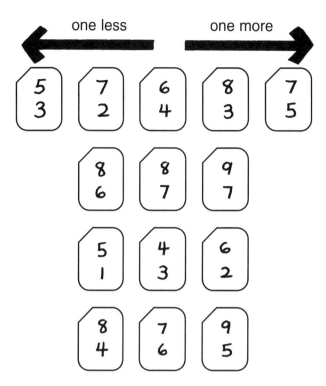

If a player can't play a card from his or her hand, then the player must draw a card from the pile until a card is drawn that can be played.

A player wins if all the cards are played from his or her hand. If neither player has put down all his or her cards, and all the cards have been drawn from the pile, the player who has the fewest cards is the winner.

ADDITION — MENTAL MATH

The commutative or order property of addition and the associative or grouping property that were introduced in earlier manipulative activities can help students do mental math.

OBJECTIVE: Students use the properties of addition in mental math.

ACTIVITY 1
Students have learned that in addition the order of two addends can be changed and the sum is still the same. This commutative or order property of addition is an important tool in learning the basic facts and in adding any numbers. Students discover that knowing one combination really is knowing two. If 9 + 6 = 15, then 6 + 9 = 15. Students should continue to model this property with manipulatives, coins and drawings.

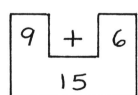

ADDITION

Another valuable tool is the associative or grouping property of addition. This property allows more than two addends to be grouped in pairs in any way.

$$7+6+4$$
$$7+(6+4)$$
$$7+10$$
$$17$$

Encourage students to group addends so that facts for 10 or any of their favorite combinations are added first.

$$20+38+80$$
$$(20+80)+38$$
$$100+38$$
$$138$$

Changing both the order and grouping of addends will enable students to add in ways that best suit them as individual thinkers.

$$2+9+3+8$$
$$(8+2)+(9+3)$$
$$10+12$$
$$22$$

Ask students to explain their choices and why they use them. Other students will benefit from hearing about different methods. The examples and discussion will convince them of the value of the properties.

ACTIVITY 2 – Human Versus Machine

This activity emphasizes the best times to use either mental math or the calculator. This is also a convincing way to show parents that using the calculator for basic facts is not efficient.

One student who represents the "human" does not use a calculator and calls out the answer as soon as he or she determines it. The other students use calculators and must enter all the digits and symbols and give the answer only when it appears on the display. Start all calculations with zero on the calculator display.

Students close their eyes as you write the exercise: 20 + 16 + 80 = ?

Then say, "Go." The "human" will call out "116" before the calculators show the sum on the display. After three or four exercises, it is obvious that the "human" is faster for certain exercises. Then give the students some exercises that are completed more quickly and accurately by the calculator than by the "human." Point out to the students that they should always look before they compute. When the mind is faster, do not use the calculator.

ADDITION — ADDING LARGER NUMBERS

Part of the problem-solving approach to mathematics involves giving students the opportunity to experiment with various ways of adding before a standard form is taught in class. Students can work in small groups and exchange ideas on how to solve the problem. Various methods can be shown to the whole class.

OBJECTIVE: Students add large numbers using different strategies.

ACTIVITY

Use the following example:
Mary has 75¢. She wants to buy a ruler for 46¢ and a large eraser for 27¢. Does she have enough money? Show how you know the answer.

Examples of some methods that students may use to solve the problem:

METHOD 1

$$\begin{array}{r} 46¢ \\ +27¢ \\ \hline +13¢ \rightarrow \text{ones} \\ +60¢ \rightarrow \text{tens} \\ \hline 73¢ \end{array}$$

METHOD 2

$$\begin{array}{r} 46 \rightarrow +4 \rightarrow 50 \\ +27 \rightarrow -4 \rightarrow 23 \\ \hline 73¢ \end{array}$$

METHOD 3

$$\begin{array}{r} 46 \\ +27 \quad \text{carry ones} \\ \hline 73¢ \end{array}$$

4 METHODS

METHOD 4

$$46¢ + 27¢$$
$$(40 + 20) + (6+7)$$
$$\underline{60 \quad + \quad 13}$$
$$73¢$$

Ask the students to explain the methods they used to solve the problem. They should feel confident that their methods are acceptable.

ADDITION

PALINDROME

A palindrome is a number that reads the same forward or backward. For example, 3113 is a number palindrome. Creating palindromes provides practice in addition of larger numbers.

Example:

```
   5 7
 + 7 5
 -----
   132

 +231
 -----
   363
```

Start with a number that is not a palindrome.

Step 1 – reverse the digits

Step 2 – add the numbers; the sum is not a palindrome.

Step 3 – reverse the digits

Step 4 – add - PRESTO! a palindrome! This is a two-step palindrome because two additions were required.

ACTIVITY

As a group activity, ask the students to find the palindromes of numbers 1 through 100. Record the number of steps it takes to make the palindrome of a number. Students should look for patterns in the number of steps it takes to make a palindrome. Why are the numbers 89 and 98 difficult numbers to find the palindrome?

Ask each student to pick three numbers and do the Palindrome Steps. On a poster or on cards, have students record the results.

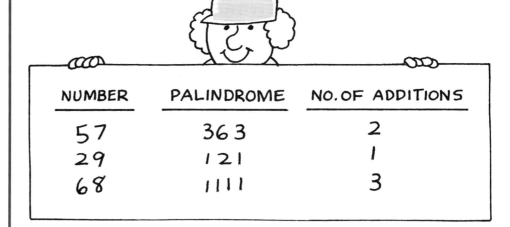

NUMBER	PALINDROME	NO. OF ADDITIONS
57	363	2
29	121	1
68	1111	3

34

SUBTRACTION

MEANING OF SUBTRACTION

The meaning of subtraction is taught most effectively through activities in which some of the objects are taken from a set to find the other part. This is the "take away" method. The "add on" model, in which objects are added on to a given part to make the total set, is also demonstrated in the following activities.

OBJECTIVE: Students model the subtraction facts by the "take away" method.

ACTIVITY

Students can use a frame to show the meaning of subtraction. Tell a story such as: "There are six cookies and Sue ate two of the cookies, how many are left?"

Ask the students to place six counters or chips in the top of the frame. Then show "take away" by pulling down two objects. The four objects are then pulled down to show the missing part. Ask the students to make up their own stories to match the "take away" action.

Students need to record each activity as it is completed in order to achieve a command of the subtraction facts.

OBJECTIVE: Students model the subtraction facts by the "add on" method.

ACTIVITY 1

Ask a comparison question that involves two amounts. "Jon colored seven squares. Beth colored three squares. How many more does Beth need to color to have the same number?" Using graph paper and crayons ask the students to model the problem. Ask the students to color seven squares in one row. Then in the row below color three squares. Count the number of squares to "add on" to the find given sum.

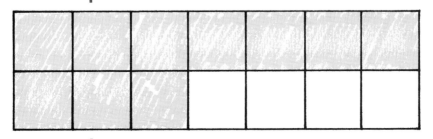

35

SUBTRACTION

ACTIVITY 2

Using Unifix Cubes or Cuisenaire rods students can model the fact 8 − 3 = ? by adding on. Ask students to arrange eight cubes or rods in one line representing the sum and three cubes or rods below. The missing part is the "add on" to make the sum. "What number is added to three to get eight?"

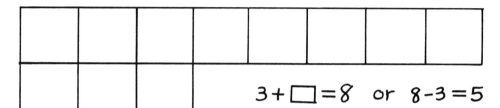

$3 + \square = 8$ or $8 - 3 = 5$

OBJECTIVE: Students, given a sum, write facts that have the given sum.

ACTIVITY

Using chips or counters, ask students to make a set of eight, and then remove chips from the set and record the fact. Each new set must start with the original sum of eight.

$8 - 1 = 7$ $8 - 2 = 6$

OBJECTIVE: Students write stories whose solution is the given fact.

ACTIVITY

Tell students that you are going to give them the "headline" and they will write a story. The "headline" is a fact such as:

$$5 + 3 = 8$$

Then ask students to write stories that have 5 + 3 = 8 as the solution. Students exchange and discuss the various stories. Instead of writing, some students could draw pictures.

SUBTRACTION

THE HILL METHOD OF SUBTRACTION FACTS 11 THROUGH 18

The subtraction facts are a major part of the mathematics program in grades one and two, however, many students in third and fourth grades still require more practice on the subtraction facts 11 through 18. One method that can be used is the hill method that involves two steps – first, make a ten and then add on a digit in the ones place.

OBJECTIVE: Students subtract using an "add on" strategy with facts 11 to 18.

ACTIVITY

Write the exercise $\begin{array}{r}15\\-6\\\hline\end{array}$ on the chalkboard. Ask the students to explain the meaning. (What number added to 6 will be 15?) The hill method is a way to do subtraction by adding. Draw a hill and write 15 and 6 on the ends and write 10 in the middle. Then ask your students: "What do you add to 6 to get 10?" (4) Write the number 4 between 6 and 10 on the hill. "What do you add to 10 to get to 15?" (10 to 15 is 5) Write the number 5 between 10 and 15 on the hill. "Now add 4 + 5." "What is the answer?"

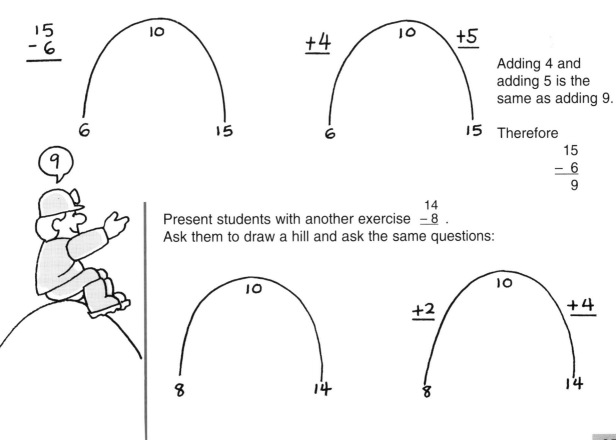

Adding 4 and adding 5 is the same as adding 9.

Therefore
$\begin{array}{r}15\\-6\\\hline 9\end{array}$

Present students with another exercise $\begin{array}{r}14\\-8\\\hline\end{array}$.
Ask them to draw a hill and ask the same questions:

37

SUBTRACTION

$$\begin{array}{r}14\\-8\\\hline 6\end{array}$$

"What do you add to 8 to make 10?" (2)
"What do you add to 10 to get to 14?" (4)
"Adding 2 and adding 4 is the same as adding 6."
"What is the answer?" or "What number added to 8 gets to 14?"

Students need to make at least four hills for subtraction to be sure they understand the concept that they add on to 10 and then add the ones. Then they should be given some subtraction facts to see if they can arrive at the answer without using the hill. To assess if the model has helped students, write some facts on the chalkboard without drawing the hill.

$$\begin{array}{r}15\\-8\\\hline\end{array}$$

"What do you add to 8 to make 10?" (2)
"What do you add to 10 to make 15?" (5)
"Add 2 and 5 and write 7 for the answer."

$$\begin{array}{r}17\\-9\\\hline\end{array}$$

"What do you add to 9 to make 10?" (1)
"What do you add to 10 to make 17?" (7)
"Add 1 and 7 and write the answer 8."

After several days of practicing the facts without the hills, the model should be firmly established in the students' minds. Ask students to share other methods they use in subtraction.

SUBTRACTION

TRIANGULAR FLASH CARDS

To better understand subtraction concepts, students need to see the relationships of the three numbers in addition and subtraction facts. This can be accomplished by using simple triangular-shaped flash cards containing the three numbers in a number family. Using heavy-duty cardstock, outline and cut out large triangles of a size that can be easily seen by all students. Using a heavy-duty marking pen, print the number facts on the cards. In printing the numbers on the cards, the sum on each card should be a different color. For example, in the family 8, 6 and 14, the 14 should be printed using a different color marking pen.

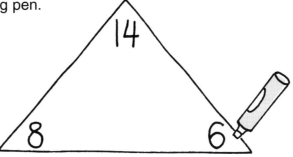

OBJECTIVE: Given three numbers in a family, the students write four facts – two addition facts and two subtraction facts.

$8 + 6 = 14$

$6 + 8 = 14$

$14 - 6 = 8$

$14 - 8 = 6$

CLASS ACTIVITY

While displaying the flash card, ask students to write the two addition and two subtraction facts. Review the order property for addition.

Continue with this activity, using the remaining prepared flash cards.

GROUP ACTIVITY

Working in pairs, students can use the triangular cards for additional practice in learning subtraction and addition facts. The first student covers one number on the flash card and the other student states the missing number. The answer is checked by revealing the covered number.

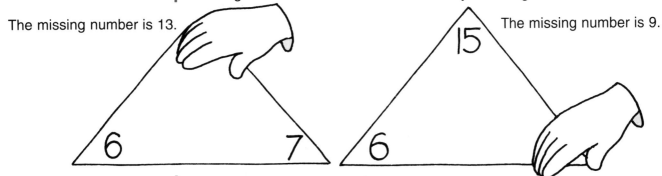

The missing number is 13.

The missing number is 9.

Some students might need to draw the number objects or use manipulatives to model the quantities. In individual practice, suggest that students work on a few number families at a time. Select those that need practice.

SUBTRACTION

SUBTRACTING TWO-DIGIT NUMBERS

In the process of subtracting large whole numbers, there are some steps to be considered. The first step is to "scan and decide." Students need to look at the ones digits to see if the numbers can be subtracted without trading.

OBJECTIVE: Students compare the digits in the ones place and decide whether or not trading is needed to subtract.

ACTIVITY

Prepare a worksheet with exercises such as those illustrated below and distribute to the students.

```
  9      2      8      0      9      1
 -2     -5     -5     -7     -3     -5
 ___    ___    ___    ___    ___    ___
 □      □      □      □      □      □
```

Ask the students to write "Y" (yes) or "N" (no) in the box under each exercise. Ask questions such as: "Can you subtract 2 ones from 9 ones?" (Y) "Can you subtract 5 ones from 2 ones?" (N) If a student answers incorrectly, use manipulative materials such as cubes or chips and ask him/her to model the example. Discuss what must be done if "No" is the response.

After your students practice "scan and decide" and have correctly responded to the single-digit problems, present two-digit problems using the same technique and questioning:

```
  4│8            6│0
 -1│3           -2│7
 ──┴──          ──┴──
   □              □
```

"Can you subtract 3 ones from 8 ones?" (Yes) "Do the subtraction."

"Can you subtract 7 ones from 0?" (No) "Just write N for no."

40

SUBTRACTION

This activity reinforces the fact that the students' first step in subtracting two-digit numbers must be to see if subtraction can be done without trading. They are learning to look at the numbers before using a pencil or calculator to do the actual calculation with whole numbers.

Review the steps in trading 1 ten for 10 ones and then complete the subtraction.

In the next step, students look at two possible answers and circle the answer they think is correct. They should be prepared to support their answers.

"Circle the answer you think is correct. Why do you think it is correct?"

If a student circles 34, he or she must show that 34 + 27 = 53. The clue for the student here is to go back and look at the ones.

Provide other examples for the student to be the "teacher" checking the work.

SUBTRACTION STRATEGIES

One of the changes made in the teaching of mathematics is to allow students to develop their own strategies to compare and to solve mathematics problems. Another change is to allow students to work in groups to compare their work and learn to listen to the thinking of others. During these activities, calculators can be used as well as paper and pencil. What is important is that students communicate to their classmates the process they used to arrive at their solution.

OBJECTIVE: Through group activities students develop their own strategies for subtraction.

ACTIVITY

Divide the students into small groups. Give each group an exercise to be solved in their own way. Tell the students that they are to work together to come up with a subtraction strategy. They should be able to explain and verify their work.

The following are examples of methods that could be employed in subtracting 19 from 37.

One method students may discover is renaming numbers to make them easier to subtract.

41

SUBTRACTION

METHOD 1

$$\begin{array}{r} 37 \rightarrow 38 \\ -19 \rightarrow -20 \\ \hline 18 \end{array}$$

Add 1 to 37 to make 38
Add 1 to 19 to make 20
Reason 1 − 1 = 0

Check: 19 + 18 = 37

METHOD 2

$$\begin{array}{r} 37 \rightarrow 37 \\ -19 \rightarrow -20 \\ \hline 17+1 \\ 18 \end{array}$$

Add 1 to 19 and subtract
Add 1 to the sum 17
to balance the 1 added

Check: 18 + 19 = 37

METHOD 3

$$\begin{array}{r} 37 \\ -19 \end{array} \rightarrow \begin{array}{r} 17-9 = \;^{+}8 \\ 20-10 = \;^{+}10 \\ \hline 18 \end{array}$$

A subtraction fact
used 1 ten,
only 2 tens left

Check: 18 + 19 = 37

METHOD 4

$$\begin{array}{r} 37 \rightarrow (20+17) \\ -19 \rightarrow -(10+9) \\ \hline 10 + 8 = 18 \end{array}$$

Renaming both numbers

In Method 3, 17 is subtracted from 37, leaving 20; 9 is subtracted from 17 which equals 8; 10 then is taken from the 20 leaving 10. The 8 and 10 equals 18.

After the students have discovered various methods of subtracting, the groups should explain their strategies to the whole class.

Put the following problems on the chalkboard and continue the activity by asking the students questions to promote more discussion of the methods, such as: "Which method would you use to solve the following problems?" "Why?"

$$\begin{array}{r} 27 \\ -\;8 \\ \hline \end{array} \quad \begin{array}{r} 98 \\ -19 \\ \hline \end{array} \quad \begin{array}{r} 40 \\ -\;6 \\ \hline \end{array} \quad \begin{array}{r} 72 \\ -37 \\ \hline \end{array}$$

SUBTRACTION

PROBLEM SOLVING — EDUCATED GUESSING

This problem-solving activity, Educated Guessing, involves the students in guessing and checking to compare values. Prepare worksheets for the students with the number sentences and record charts.

OBJECTIVE: Given an open mathematical sentence, students make guesses, record the data and compare results.

ACTIVITY

Write the following number sentence on the chalkboard: (☐ + 4) − 7 = 23. Then ask the students to guess what number will make this sentence true.

Try 10 (10 + 4) − 7 = 7
7 is less than 23
Try 20 (20 + 4) − 7 = 17

NUMBER GUESSED	(☐+4)−7	TRUE?
10	7	NO
20	17	NO
26	23	YES

After the second guess of 20, point out that the answer (17) is less than the answer we need, which is 23. Ask the students how to decide the next guess. Students should look at each guess and find a pattern in the outcome. Students will make different numbers of guesses to find the correct values.

Distribute the worksheets to the students. "Find the number that will make each sentence true. Record your guesses and the outcome in the table. Look for a pattern."

(☐ + 5) − 7 = 38

NUMBER GUESSED	(☐+5)−7	TRUE?

Number is _____
Check: (__ + 5) − 7 = 38

(△ + 30) − 12 = 57

NUMBER GUESSED	(△+30)−12	TRUE?

Number is _____
Check: (__ + 30) − 12 = 57

SUBTRACTION

$$(\square - 8) + 12 = 43$$

NUMBER GUESSED	(□−8)+12	TRUE?

Number is ____
Check: (__ − 8) + 12 = 43

$$23 + (\triangle - 8) = 44$$

NUMBER GUESSED	23+(△−8)	TRUE?

Number is ____
Check: 23 + (__ − 8) = 44

Ask the students to make up number sentences to challenge each other.

MISSING DIGITS — I

Students can be given problems that cause them to rethink the operations. One such activity is Missing Digits, in which students can work cooperatively to solve problems and develop strategies.

OBJECTIVE: Students work cooperatively to solve problems with missing digits.

ACTIVITY
Write the following problem on the chalkboard:

```
  2 ? 1
- ? 0 ?
-------
    1 8
```

After the students have investigated the problem, ask them to share their strategies.

A trade of one 10 must be made to change the 1 to 11 in order to get the result of 8 ones. Then we know the missing digit under 1 is 3. The digit between 2 and 1 must be 2 because we needed to trade one ten to make 11 ones and 1 − 0 = 1. The last missing number is 2.

Prepare similar problems for your students. Ask them to find the missing digits and explain how they found them.

> If the students need a clue, suggest they start with the ones.

Example:

```
  6 ? 8         4,3 ? 5
- ? 3 ?         -  ? 5 ?
-------         ---------
  3 7 9         ? 5 3 8
```

SUBTRACTION

MISSING DIGITS — II

In this activity students will find the value of letters that represent the digits. (NOTE: The letters have different values in the different problems.) Two letters together such as CD represent a two-place number, and the same letter repeated, such as BB, represents a two-place number with the same digit in each place.

OBJECTIVE: Students solve problems in which letters represent missing digits.

ACTIVITY

Find the values of the letters. They will have different values in different problems. Work through one of the problems with the students eliciting responses from the students through questioning. Ask students what strategies they will use to solve these problems.

```
   A E        B 4        A B        A A        7 E
  - 7        - D        -2 7       -A 2       -E 0
  ----       ----       ----       ----       ----
   4 5        5 6        5 9         4         1 4

  A is ___   B is ___   A is ___   A is ___   E is ___
  E is ___   D is ___   B is ___              D is ___
```

45

SUBTRACTION

FRAME ARITHMETIC

The activity of frame arithmetic is designed to reinforce the basic addition and subtraction facts and to consider two conditions in problem solving. Problems in frame arithmetic usually involve problem-solving techniques of trial and improve. (Trial and improve is much better than trial and error.) Prepare worksheets for the students with samples of frame arithmetic problems.

OBJECTIVE: Students find pairs of numbers that satisfy both addition and subtraction equations.

Begin with several examples of pairs of numbers with the same sum. Write the sentences to illustrate the sum and the difference.

$$\square = 7 \quad \triangle = 6 \qquad \square = 9 \quad \triangle = 4$$

$$\boxed{7} + \triangle_{6} = 13 \qquad \boxed{9} + \triangle_{4} = 13$$

$$\boxed{7} - \triangle_{6} = 1 \qquad \boxed{9} - \triangle_{4} = 5$$

ACTIVITY

Explain to your students that in frame arithmetic the same number is used in each square and each triangle. Present the following problem-solving exercise to your students.

Ask the students: "What two numbers when added together have the sum of 14, but when subtracted have a difference of 4?"

Students will try 7 and 7 for a sum of 14, but the difference of 7 – 7 is 0. Then they could try 8 + 6 for a sum of 14, but the difference of 8 – 6 is 2. They might try 9 + 5 for a sum of 14 and then see the difference of 9 – 5 is 4.

Using two equations in frame arithmetic not only gives valuable practice with addition and subtraction facts, but also provides a problem-solving strategy of "trial and improve." Students also need to learn that many tries are needed. Sometimes the first try gives special clues.

Organizing the trials in a table could help some students consider the possible values.

SUBTRACTION

OTHER METHODS OF ADDITION AND SUBTRACTION

There are two different ways to add and subtract whole numbers used in other countries. After practicing these methods many students find the new ways easier than the standard algorithm they have been taught.

OBJECTIVE: Students add whole numbers by the Check Method.

ACTIVITY

In England, the Check Method of adding is taught. Each time you add two one-digit numbers, a check (✔) is made for the ten. The number of checks in each column is written at top of the column to the left.

```
    6          ²6
    7          7₃
    8          8₁
  + 5        + 5
             ___
              26
```

6 + 7 = 13 check (✔) for ten write 3
3 + 8 = 11 check (✔) for ten write 1
1 + 5 = 6
For 2 checks in ones column, write 2 on top of the tens column

```
   684              2  2
  3178           1  6  8₀ 4
   243           3  1  7  8₂
 +1678           2₁ 4₁ 3
                +1  6  7  8
                _____
                 5  7  8  3
```

4 + 8 = 12 check for ten
2 + 3 + 8 = 13 check for ten write 3, "carry" two tens to next column
2 + 8 = 10 check ten
0 + 7 + 4 = 11 check ten
1 + 7 = 8 write 8, "carry" 2 checks
2 + 6 + 1 + 2 = 11 check ten
1 + 6 = 7 write 7, "carry" 1 check
1 + 3 + 1 = 5
write 5

The Check Method eliminates two trouble spots in the algorithm used in the United States. One trouble spot is adding a two-digit number to one-digit seen number. The other is carrying the wrong number to the next column.

```
      4 8    → 14 is unseen
    2 7 6
  +   3 7
```

47

SUBTRACTION

OBJECTIVE: Students subtract by the Compensation Method.

ACTIVITY

Many countries teach the Compensation Method of subtraction. The method was used until 1955 in the state of New York and Chicago and Los Angeles.

Example:

$$\begin{array}{r} 600\overset{1}{0} \\ -\ 27\underset{5}{4}8 \\ \hline 2 \end{array}$$

Can't subtract 8 from 0.
Add 10 to the sum.
Then change 4 in tens place to a 5.

$$\begin{array}{r} 60\overset{1}{0}\overset{1}{0} \\ -\ 2\ 7_{8}\underset{5}{4}8 \\ \hline 5\ 2 \end{array}$$

Can't subtract 5 from 0.
Add 10 to the sum.
Then change 7 to 8 in the hundreds place.

$$\begin{array}{r} 6\overset{1}{0}\overset{1}{0}\overset{1}{0} \\ -\ 2_{3}7_{8}\underset{5}{4}8 \\ \hline 2\ 5\ 2 \end{array}$$

Can't subtract 8 from 0.
Add 10 to the sum.
Change 2 to 3 in the thousands place.
Subtract 3 from 6.

Example:

$$\begin{array}{r} 5\overset{1}{1}\overset{1}{4}\overset{1}{7} \\ -\ 2,8,6,9 \\ \hline 2\ 2\ 7\ 8 \end{array}$$

Instead of changing the number in the addend, students can write a 1 to be added to the number when 10 is added.

The adding of 1 instead of crossing out a number makes it easier to check the problem.

48

MULTIPLICATION/DIVISION

Chapter 3

BACKGROUND

The NCTM Standards stress the importance of understanding the concepts of the operation of numbers. Memorizing rules for computation without understanding the fundamental concepts has caused many students to feel they could not do mathematics. Adults who dislike mathematics often say their dislike is due to never understanding what they were doing as students.

There are three important stages in developing understanding while learning mathematics:
- the concrete stage
- the semi-concrete or representational stage
- the abstract stage

In the initial stages of learning a new concept students manipulate concrete materials such as chips, Unifix Cubes, Cuisenaire rods, pattern blocks and attribute blocks to construct the meaning of operations and mathematical ideas. The development of these concepts continues with pictorial models or drawings that represent the key ideas. The abstract stage involves attachment of symbols to quantities, operations or relationships.

You must be ready to move the instruction and lessons from stage to stage as your students indicate their readiness to do so. All stages should be available to students for all concepts to insure that they are building on understanding.

Students who understand what multiplication means are on surer ground than students who simply memorize the "times table." As with addition and substraction, one way to help students understand the concept of multiplication is to offer activities using manipulatives.

It is also important for students to understand the properties or rules of multiplication:

- The commutative property, which means that the factors can switch places and the product stays the same.

- The associative property, which means that when there are more than two factors, you can group them in any order and the product stays the same.

- The identity element of the number 1. This means that when 1 is a factor, the product is the same as the other factor.

- The property of zero. This means that when zero is a factor, the product is zero.

Division is the inverse operation of multiplication and is related to the repeated subtraction of an equal amount. Many students find it particularly hard to perform the division algorithm. It is important to help students develop an understanding of the basic division concept by offering them the use of math manipulatives and guided activities. Students then can create models of division on which they can base their calculations.

MULTIPLICATION

MANIPULATIVE/PICTORIAL MODELS

Through the use of manipulatives and drawings, the following student activities prepare students for the abstract stages of multiplication and division.

OBJECTIVE: Students use manipulatives or pictorial models to represent multiplication.

ACTIVITY 1

Ask students to arrange 14 counters in 2 groups of 7. Discuss how this arrangement could be labeled 7 + 7 or 2 x 7. Then ask students to make groups of 2 and label the new arrangement. Continue this activity with different quantities. Students also can work in pairs — one student makes the groups and the other student states the two number names.

7 + 7 = 14
2 SETS of 7 = 14
2 x 7 = 14

2 + 2 + 2 + 2 + 2 + 2 + 2 = 14
7 SETS OF 2   = 14
7 X 2   = 14

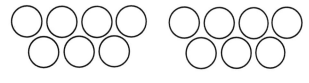

3 x 5 = 15

5 x 3 = 15

Making rectangular arrays with the same number of objects in each row is the next stage in investigating the factors. The array will lead students to understand that the order of the factors does not change the product even though the picture looks different.

In multiplication, the order of the factors can be switched and the product will still be the same. This order property is a valuable rule and is called the commutative property of multiplication. Students readily see that 3 x 5 is 5 + 5 + 5 and 5 x 3 is 3 + 3 + 3 + 3 + 3.

Ask students to draw arrays of 6 fours or 6 x 4 and compare them to arrays of 4 sixes or 4 x 6. Ask students to work with a partner; one gives a multiplication fact and the other draws and labels it.

MULTIPLICATION

ACTIVITY 2
Use paper with squares to show the area model of the meaning of multiplication.

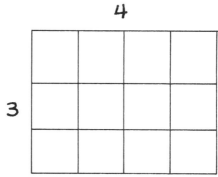

3 rows of 4 squares

SHOWS 3 × 4 = 12

ACTIVITY 3
Use paper with squares to show that the order of the factors will not change the product, but the pictures are different. Discuss why this order property (commutative) of multiplication is useful.

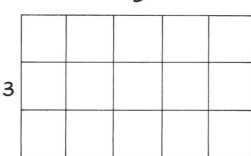

FRAME ARITHMETIC

Frame arithmetic provides practice with the abstract stage of using numbers and operations in number sentences. Suggest that students can always check their work by using concrete pictures or objects.

When the frames are alike, the same number is used for each frame.

OBJECTIVE: Students use frame arithmetic to practice the abstract stage of using symbols for numbers and operations.

MULTIPLICATION

ACTIVITY
Ask the students, "What numbers make this sentence true?"

☐ + ☐ + ☐ = 12

4 + 4 + 4 = 12

Three fours equals 12, therefore, 3 x 4 = 12.

Now, "What numbers make this sentence true?"

☐ + ☐ + ☐ + ☐ = 12

3 + 3 + 3 + 3 = 12

Four threes equals 12, therefore, 4 x 3 = 12.

Students recognize that 3 x 4 and 4 x 3 have different models with equal addends, but both facts have the product of 12. Again, the order property is used as a helpful tool.

EXTENSION
Students can make "product books" with pictures of things that come in sets of 2, 3, 4, 5, 6, 7, 8 or 9. Each illustration should have a sentence description for the selected fact. The multiplication table for each number can also be on the page.

Example:
    FOUR
    Chairs have 4 legs.
    There are 8 chairs in the circle,
    so there are 32 legs.
    4 + 4 + 4 + 4 + 4 + 4 + 4 + 4 = 32
    8 x 4 = 32

    1 x 4 = 4
    2 x 4 = 8
    3 x 4 = 12 etc.

MULTIPLICATION

ASSESSING THE MEANING OF MULTIPLICATION
One of the methods of assessing the meaning of multiplication is to list numbers to be added in a column. Students then are asked to decide whether the problem can be written in multiplication form.

Example:
"Which of the columns of numbers can be rewritten as a multiplication problem?"

```
    6           7          13         3
    6           8          13         7
  + 6         + 3        + 13         8
    6           9          13         5
  ─────       ─────      ─────      ─────
  YES 4×6      NO        YES 5×13     NO
```

Another way to assess the students' understanding of the meaning of multiplication is to give students a problem such as 4 x 23 and ask how they can find the product. Students could respond by adding 23 + 23 + 23 + 23. This response shows that students understand the connection between equal addends and multiplication.

THE PROPERTIES OF MULTIPLICATION

The property of zero is an important property of mathematics. In multiplication, if zero is one of the factors, the product will always be zero. There are 19 multiplication facts with zero as one of the factors.

The identity element of multiplication is one (1). One (1) times any number will result in a product that is the number of the other factor. There are 17 multiplication facts that have one as a factor.

OBJECTIVE: Students use the properties of multiplication to learn the multiplication facts.

ACTIVITY
Help your students understand the properties of multiplication with the following examples.

Example:

$6 \times 0 = 0$

because 6 x 0 means 0 + 0 + 0 + 0 + 0 + 0. Also 0 x 6 = 0 because of the commutative property. Using a calculator, students can see that any number multiplied by zero is zero.

Example:

$5 \times 1 = 5$

because 5 x 1 means 1 + 1 + 1 + 1 + 1 and 1 x 5 = 5 because one five is five.

53

MULTIPLICATION

THE SPECIAL NUMBER 9

The number that is one less than the base in a number system is a special number. In the base ten system, the number 9 is special.

OBJECTIVE: Students identify patterns in the table of 9 facts.

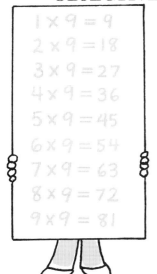

ACTIVITY

Ask the students to make a table of the 9 facts and then search for patterns in the table.

- The digits in each product add to 9.

- The tens digit of the product is one less than the factor.

    $7 \times 9 = 63$

- In the product column, the tens digits increase and the ones digits decrease.

SIX ADDITIONAL FACTS

After students know the facts of 2, 3, 4, 5 and 9, the properties of 0 and 1 and use the commutative property of multiplication, there are only six facts to learn. They are 6 x 6, 6 x 7, 6 x 8, 7 x 8, 8 x 8 and 7 x 7.

OBJECTIVE: Students learn the facts of 6, 7 and 8.

ACTIVITY

In this activity, have students decide which facts they need to model, draw and practice.

Remind students to work on only a few facts at a time.

$6 \times 6 =$

$6 \times 7 =$

$6 \times 8 =$

$7 \times 8 =$

$8 \times 8 =$

$7 \times 7 =$

MULTIPLICATION

WRITING COMPLETE EQUATIONS

Students learn the multiplication facts best when they write the whole fact. When given the fact, the students write only the product, but when given the product students write the whole fact.

OBJECTIVE: Students write complete equations to reinforce the multiplication facts.

ACTIVITY

Using 1" graph paper, the products of the multiplication tables (multiples) 2 through 9 are written in the squares. This is the same as counting by 2s, 3s, 4s, etc.

Example: The Table of Three or Multiples of Three

| 3 | 6 | 9 | 12 | 15 | 18 | 21 | 24 | 27 | 30 |

Then ask students to write all the facts for the table of three:

$1 \times 3 = 3 \quad 2 \times 3 = 6 \quad 3 \times 3 = 9 \quad 4 \times 3 = 12 \quad 5 \times 3 = 15$

$6 \times 3 = 18 \quad 7 \times 3 = 21 \quad 8 \times 3 = 24 \quad 9 \times 3 = 27 \quad 10 \times 3 = 30$

These table strips can also be used for oral activities. For example, ask students to point to the number 18 and state the fact. "How many 3s in 18?" or "Give me a '3' fact."
        Point to 27,
        "How many 3s in 27?"

When needed, have students make arrays for their own identified "hard" facts. Ask students to share any methods they use for remembering certain facts.

MULTIPLICATION

TRIANGULAR FLASH CARDS

The division facts can be learned after students know the multiplication facts. The division facts are actually the missing factor in multiplication facts.

Example: ? x 6 = 48   or   48 ÷ 6 = ?   How many 6s in 48?

OBJECTIVE: Students use multiplication facts to find the related division facts.

ACTIVITY

Using the Triangular Flash Cards, ask students to write four related facts. In creating the flash cards, the product should be printed in a color different from the factors.

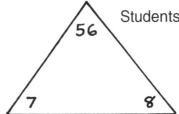

Students write:
7 x 8 = 56
8 x 7 = 56
56 ÷ 8 = 7
56 ÷ 7 = 8

Writing the four facts will reinforce for your students the relationship of multiplication and division.

MULTIPLY AND ADD

For many students, the trouble spot in compound multiplication appears when it is necessary to multiply and then add. Research shows that when students multiply many digits, about 70% of the errors occur when they have to multiply and then add.

OBJECTIVE: Students practice the sequence — multiply and add.

ACTIVITY

Use an example such as the following to illustrate multiplying and adding:

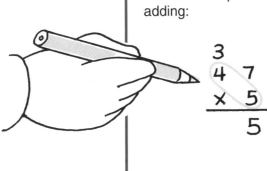

Multiply 5 x 7 ones; multiply 5 x 4 tens, then add 3 tens.

20 tens + 3 tens = 23 tens

23 tens plus 5 ones = 235

MULTIPLICATION

HIGH OR LOW GAME

Practicing multiplication with addition in a game-like activity helps students understand multiplication concepts and remember the multiplication facts. The game of High and Low encourages this practice through a group activity that is non-competitive in the sense that game skills are not required, but rather points are earned through the chance dealing of the cards. High and Low can be played by groups of 4 to 6 students. Each group will need a deck of playing cards with all face cards and Jokers removed.

OBJECTIVE: Students practice multiplying two numbers and adding a number to the product.

ACTIVITY

One player in each group deals out two cards facedown and one card faceup to each player.

Rules of Play: Players do not pick up their cards. The dealer states either "high" or "low" and then turns over his or her two facedown cards. The dealer multiplies the two cards and then adds the number on the third card.

Example:  SCORE: (4 × 8) + 7 = 39

The remaining players now turn over their cards, multiply the numbers and add the third number to get a total score.

If the score of a player is more than the dealer's score after the dealer has said "high," that player earns 10 points. If more than one player scores higher than the dealer's score, each one earns 10 points. If no one has a score higher than the dealer's score, the dealer earns 10 points.

If the dealer has said "low," then the other players would have to have scores less than 39 to earn 10 points. If all the other scores are higher than the dealer's score, the dealer earns 10 points. If more than one player has a score lower than the dealer's score, each player earns 10 points.

After the first round of play, another player shuffles the cards and deals the cards as described.

The game is over after each player has had an opportunity to be the dealer. The player with the highest score is the winner.

57

MULTIPLICATION

MULTIPLYING A MANY-DIGIT FACTOR BY A ONE-DIGIT FACTOR

Proficiency in multiplying by a 1-digit factor is important for finding both exact answers and estimates either with paper and pencil or by mental computation. There are several steps used to build understanding and efficiency in the multiplication of a many-digit factor by a 1-digit factor.

OBJECTIVE: Students multiply many-digit factors by 1-digit factors.

ACTIVITY

Demonstrate for your students many-digit multiplication using the following examples:

Step 1
Multiply a multiple of 10, 100 or 1000 by a 1-digit factor.

```
  4 tens        40         6 hundreds      600
  × 8          × 8         × 4            × 4
  ──────       ────        ──────────      ────
  32 tens      320         24 hundreds    2400
```

As students think of 40 as 4 tens and 600 as 6 hundreds, they are using basic facts to find the products that are also multiples of 10, 100 or 1000.

Step 2
Determine the number of zeros in the product.

```
   400         500         7,000         8,000
   × 7         × 6         × 8           × 5
   ────        ────        ──────        ──────
  2800        3000        56,000        40,000
```

Underline the first digit and multiply it by the other factor using basic facts. Then write the appropriate number of zeros in the product.

Ask the students to discuss how to decide the correct number of zeros in the product. Make sure they consider the exception when the basic fact product ends in a zero. Using place-value names as done in Step 1 is also helpful in avoiding errors.

Provide practice of both of these steps written in horizontal form.

Example: 9 × 3000
          7 × 600
          8 × 5000

58

MULTIPLICATION

Step 3
Use the distributive property of multiplication over addition.

Using graph paper, illustrate how a two-digit number can be written as an indicated sum. Ask students to outline a 3 by 14 array, that is, 3 rows of 14 squares. Then separate the rows of the first 10 squares from the rows of 4 squares with a bold line. (Cutting out the rectangles is also very effective.) Students now see 3 rows of 10 and 3 rows of 4 are the same as 3 rows of 14. This shows that 3 x 14 can be thought of as (3 x 10) plus (3 x 4).

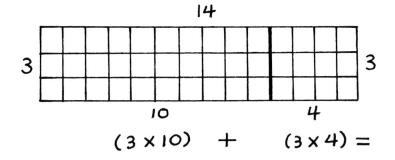

$3 \times 14 = ?$
$3 \times 14 = 3 \times (10 + 4)$
$3 \times 14 = (3 \times 10) + (3 + 4)$
$3 \times 14 = 30 + 12$
$3 \times 14 = 42$

Using the distributive property, students learn to mentally multiply two-digit numbers by a one-digit number. Students will do the work mentally as they see that multiplying by ten is easy and the other multiplication is a basic fact.

MULTIPLICATION

Step 4
Multiply 2- and 3-digit numbers by a 1-digit factor.

Examples:
$$\begin{array}{r}23\\ \times 6\\ \hline\end{array} \qquad \begin{array}{r}328\\ \times 7\\ \hline\end{array}$$

You know:
$$\begin{array}{r}20\\ \times 6\\ \hline 120\end{array} \quad \text{and} \quad \begin{array}{r}3\\ \times 6\\ \hline 18\end{array} \quad \text{then} \quad \begin{array}{r}23\\ \times 6\\ \hline 18\\ 120\\ \hline 138\end{array} \quad \begin{array}{l}\longleftarrow 6 \times 3\\ \longleftarrow 6 \times 20\end{array}$$

You know:
$$\begin{array}{r}300\\ \times 7\\ \hline 2,100\end{array} \quad \begin{array}{r}20\\ \times 7\\ \hline 140\end{array} \quad \begin{array}{r}8\\ \times 7\\ \hline 56\end{array} \quad \begin{array}{r}328\\ \times 7\\ \hline 56\\ 140\\ 2100\\ \hline 2296\end{array} \quad \begin{array}{l}\longleftarrow 7 \times 8\\ \longleftarrow 7 \times 20\\ \longleftarrow 7 \times 300\end{array}$$

After doing the long form, students can learn the short form. The trouble spot in the short form is to multiply and add.

$$\begin{array}{r}328\\ \times 7\\ \hline\end{array} \longrightarrow \begin{array}{r}300\\ \times 7\\ \hline\end{array} = 7 \times 3 \text{ hundreds} = 21 \text{ hundreds}$$

Often the equation form is easier to use. Short form example. As the short form is being used, students should be asked to estimate the product before they multiply.
We estimate to check to see if our products are reasonable. Ask students to explain their estimation methods.

$$\begin{array}{r}328\\ \times 7\\ \hline 2,296\end{array}$$

$$\begin{array}{r}415\\ \times 6\\ \hline\end{array} \rightarrow \begin{array}{r}\underline{}\\ \times 6\\ \hline\end{array} \qquad \begin{array}{r}298\\ \times 8\\ \hline\end{array} \rightarrow \begin{array}{r}\underline{}\\ \times 8\\ \hline\end{array} \qquad \begin{array}{r}526\\ \times 7\\ \hline\end{array} \rightarrow \begin{array}{r}\underline{}\\ \times 7\\ \hline\end{array}$$

      estimate   estimate   estimate

MULTIPLICATION

ASSESSING THE MEANING OF MULTIPLICATION

Determining the missing factor helps assess the students' understanding of the meaning of multiplication with multiples of hundred and thousands. To fill in the missing number, the students must look at the product and decide which factor will make the product. This reverse activity requires more thinking than multiplying two factors.

OBJECTIVE: Students identify the missing factor in multiplication with multiples of hundreds and thousands.

ACTIVITY 1
Present students with exercises such as the following:

4 × ___ = 28   6 × ___ = 4,200
4 × ___ = 280   6 × ___ = 42,000
4 × ___ = 2,800   30 × ___ = 1,200
4 × ___ = 28,000   30 × ___ = 12,000

ACTIVITY 2
Students can work in groups with each student using a calculator. Challenge the group to use only numbers 1, 2, 3 and 4 to find the greatest product and the least product. Ask the students to explain where they placed the numbers and why. Each student calculates the product and the students compare answers. Students can select four other one-digit numbers and then follow the same rules.

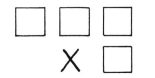

Use 1, 2, 3, 4 only once in the squares.

MULTIPLICATION

CASTING OUT NINES

One method of checking multiplication is called casting out nines.

Step 1: Add the single digit number until there is a one-digit number.

64325 → 20 → 2

6 + 4 = 10 + 3 = 13 + 2 = 15 + 5 = 20   2 + 0 = 2

173482 → 25 → 7

36459 → 27 → 9

Step 2: Cast out all numbers that add up to 9 and then add the remaining numbers.

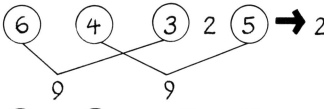

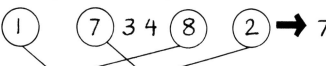

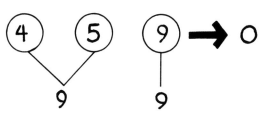

```
    4  6  2   →  12  →  3     factor
  x    3  7   →  10  →  1 x   factor
    3  2  3  4        ③      product
 1  3  8  6           ↕      checks
 1  7  0  9  4   → 12  → ③   product
```

In casting out nines, you are finding the remainder of each number when the number is divided by 9.

64325 ÷ 9   the remainder is 2

173482 ÷ 9   the remainder is 7

36459 ÷ 9   the remainder is 0

DIVISION

DIVISION READINESS

The preparation for division begins in primary grades as students make sets of a given number and make a given number of sets. This leads to the different meanings of the quotient.

OBJECTIVE: Students represent different groupings with manipulatives in a Sorting Game.

ACTIVITY 1

Prepare record sheets as illustrated and provide the students with counters such as cubes, buttons or disks. Ask the students to write the numbers 2 to 10 in the first column.

Begin with an example of 20 counters. Ask the students to write 20 in the box at the top of the record sheet. The students are to sort the counters into the given number of equal groups, determine how many counters in each group and the number of leftovers. Be sure to record zero if there are no leftovers.

Next working in pairs, the students pick up two handfuls of counters, write the number in the box at the top of the record sheet, separate the counters into the groups and record each activity.

20		
Number of Equal Groups	How many in Each Group?	Leftover
2	10	0
3	6	2
4	5	0
5		
6		
7		
8		
9		
10		

ACTIVITY 2

Using the record sheets and counters from Activity 1, ask the students to list the numbers 2 to 10 in the second column. Begin with an example of 15. Students are to make as many groups of the given number as possible. Note that the partitioning is different in this activity in which students find the number of groups.

Students can work in pairs or individually to form the groups after taking a set of counters.

Use these activities as center activities so that students regularly play the Sorting Game. Vary the directions for finding the number of groups or the number in each group.

15		
Number of Equal Groups	How many in Each Group?	Leftover
7	2	1
	3	
	4	
	5	
	6	
	7	
	8	
	9	
	10	

DIVISION

As students are ready to make the transition to the abstract concept of division, ask the students to write the division and label the quotient.

"If you put 18 cans of juice into six-packs, how many packs will you get?"

Number of groups	How many in each?	Leftover	
3	6	0	18 ÷ 6 = 3 packs

"You have 17 pennies. If you put 5 pennies in each box, how many boxes will you fill?"

17 ÷ 5 = 3 boxes and 2 pennies leftover

DIVISION MODELS

Students need to draw pictures of a division problem and then record the action. One should always record action and interpret the quotient.

OBJECTIVE: Students use manipulatives or drawings to create models for the meaning of division.

ACTIVITY
Show and model these two meanings of division:

Finding the number in equal groups:
How would you divide 23 pennies among 5 children? How many pennies would each child receive? Are there any pennies leftover? The students should draw:

$$\begin{array}{r} 4\,r\,3 \\ 5\overline{)23} \end{array}$$

Answer: Each child receives 4 pennies and there are 3 pennies leftover.

Finding the number of equal groups:
You have a bucket with 26 crayons. How many boxes with 8 crayons in each can you fill? Are there crayons leftover?

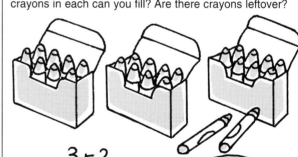

$$\begin{array}{r} 3\,r\,2 \\ 8\overline{)26} \end{array}$$

THERE ARE 3 BOXES WITH 2 CRAYONS LEFT OVER.

DIVISION

It is important that some exercises have remainders other than zero. Help students connect the record with the drawings. Have students draw pictures for examples of both types. After drawing pictures for the problems, the students start to think of the fact that has a product nearest to the dividend.

Example: 6)31 6 × ? is nearest to 31

Think:
What number ×6 is close to 31?

4 × 6 = 24
5 × 6 = 30
6 × 6 = 36

"5 × 6 is closest to 31."

      ○
```
     5 r 1
  6)31
   -30   ← 6 × 5
     1
```

DIVISION ASSESSMENT

Before students divide large numbers, you need to assess the level of understanding of division with about three exercises. If students can find the missing numbers in each exercise, this demonstrates their understanding of the quotient and remainder in division.

OBJECTIVE: Students find the missing divisor, dividend or quotient in a division example.

```
     8 r 1              7 r 4              ? r ?
  5) ?              ?) 46              8) 59
```

65

MEASUREMENT
Chapter 4

BACKGROUND

It is important that teachers create happenings in which students are involved in measuring objects to actively learn the concepts and skills of measurement. After many experiences in estimating and measuring, students learn which unit of measurement is most appropriate for the object to be measured. For example, to measure the length of a hall, the unit of measure would usually be yards or meters not inches or centimeters.

Most basic textbooks include both metric and English units. Metric measures and rulers are often easier for students to read because the system is based on units of ten. Therefore, it usually is best to start measurement activities with metric units. Later the English units can be learned with similar activities. Conversion from metric units to English units is not needed.

At the primary level linear measure is a major objective. Area and volume measurements are introduced at a readiness stage, and more development of area and volume is done in the later grades.

Estimation is an important skill. Students learn how to make good estimations after they have had many experiences of measuring and are becoming familiar with standard units.

The various skills needed for linear measurement are learned in the primary grades. Non-standard units are used as the model to introduce the skills and the concepts.

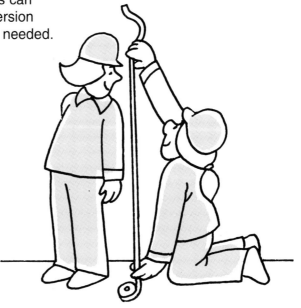

MEASUREMENT

NON-STANDARD UNITS

Measuring objects with non-standard units such as hands, feet, pencils, crayons, etc. prepares students for using standard units of linear measurement. Measurement skills are learned by example and experience.

OBJECTIVE: Students develop skills of linear measurement with non-standard units.

ACTIVITY

Students use a non-standard unit such as their hands to measure the length of a table in front of the room.

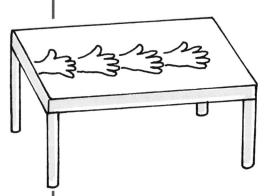

"Is this the way to measure?" Yes, start the measurement at the end of the table with no space between the hands.

"Is this the way to measure?" No, the unit of measure did not start at the end of the table, and there are spaces between the hands.

67

MEASUREMENT

OBJECTIVE: Students investigate the relationship between the size of a unit and the number of units needed to measure a given object.

ACTIVITY

Ask the students to measure the length of an object using two different size units. They could use paper clips and crayons to measure the same object, such as a large book.

Students discover that the longer the unit of measure the fewer number of units are used, and the smaller the unit of measure, the greater number of units are used.
Provide regular opportunities to do this activity.

TELLING TIME

Telling time to the nearest five minutes is an objective of measurement in the primary grades.

OBJECTIVE: Students learn to mentally visualize the numbers on the clock.

ACTIVITY

On a large clock in the classroom, cover all of the numbers except 12 with Post-It Notes with letters written on them. Point to a letter and ask: "What number is under T? B? D? F? etc." The students respond and check answers by lifting the Post-It.

The students are creating a mental clock to know where the numbers are situated without seeing them.

Make a number line from 0 to 60. Outline the multiples of 5. Use this as a model for minutes in 1 hour and practice counting by fives and identifying the numbers between the multiples.

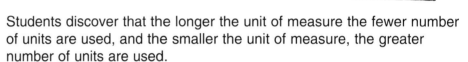

68

MEASUREMENT

OBJECTIVE: Students learn where the multiples of five are located as minutes on the clock and their sequence.

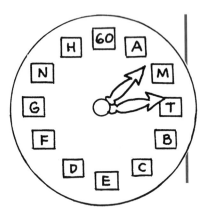

ACTIVITY
On a large clock in the classroom, put the labels of the minutes over the hours. Cover all of the numbers except 60 with Post-It Notes with letters on them as in the previous activity in random order.

Ask the student to say: 5, 10, 15, 20, etc. to 60. Then ask: "What multiple of 5 is under T? E? F?" Uncover several of the multiples as references if students need the clues.

Vary the oral practice counting the minutes or hours.

LINEAR MEASUREMENT

A major part of measurement in the primary grades is linear measure. Students learn to read a ruler and a tape measure, label the units and measure objects. Then after a great many experiences of measuring to a given unit, students can estimate length using the given unit.

OBJECTIVE: Students create a mental model or benchmark of the meter through physical and concrete experience.

ACTIVITY 1
Each student determines the relationship of the length of a meter to his or her own height. The student stands up and places one end of a meter stick on the floor and finds where the other end of the meter stick touches his or her body. Each student uses this distance as a model of a meter.

ACTIVITY 2
Working in pairs and given a heavy string one meter long, the students find objects in the room that are longer, shorter or the same as the length of a meter. Students record the names of objects in a table such as the one illustrated and then talk about and compare their tables.

The students also can take the meter string home to find objects that are less, more or the same length as a meter.

Longer than a meter	A meter	Shorter than a meter

MEASUREMENT

ACTIVITY 3
To form a benchmark for estimating long distances, students lay four meter sticks end to end from one wall toward the opposite wall. Using the length of the four meters as a benchmark, each student gives an estimate of the length of the room in meters. All estimates can be recorded on the board. Meter sticks are laid out across the room and are counted. "Who made the best estimate?" Repeating this activity by finding other lengths until estimates come closer to the actual lengths will improve students' estimating skills.

OBJECTIVE: Students measure small objects with a centimeter ruler or centimeter tape.

ACTIVITY 1
Using a 30 centimeter ruler or measuring tape and working in pairs, the students estimate and measure parts of their bodies and then record the measurements. Students compare their measurements to see if most of the measurements are about the same. For example, most ears are 6 cm. The length of the biggest smile is the most fun. On a large poster paper, students label drawings of these parts with the measurements and call the poster "Metric Me."

	Estimate	Measurement
Length of your longest finger	_____ cm	_____ cm
Length of your ear	_____ cm	_____ cm
Length of your hand from longest finger to wrist	_____ cm	_____ cm
Length from elbow to wrist	_____ cm	_____ cm
Length of your foot (no shoe)	_____ cm	_____ cm
Length of your longest smile	_____ cm	_____ cm

MEASUREMENT

ACTIVITY 2

"Distance Around" is an activity where students measure the sides of different shapes and then find the total of the measures. This activity develops readiness for perimeter. This is a good activity for students to use calculators for the addition.

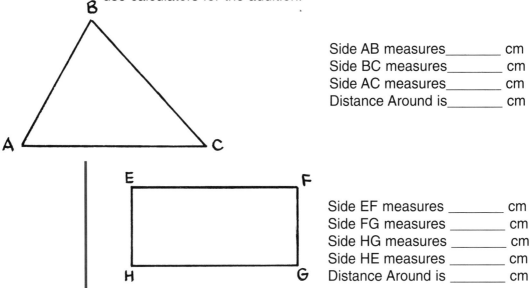

Side AB measures _____ cm
Side BC measures _____ cm
Side AC measures _____ cm
Distance Around is _____ cm

Side EF measures _____ cm
Side FG measures _____ cm
Side HG measures _____ cm
Side HE measures _____ cm
Distance Around is _____ cm

OBJECTIVE: Students compare measurements of parts of the body in ratios.

ACTIVITY 1

"What Shape Are You?" is an activity in which students compare their heights with their reaches. A reach is the length of both arms outstretched. A string is cut to match the height of each student. This string is used to measure and compare the reach. A poster is made for students to record their names in the geometric shapes. Square shapes have equal heights and reaches. Tall rectangle shapes have heights which are longer than reaches, and wide rectangle shapes have longer reaches than heights.

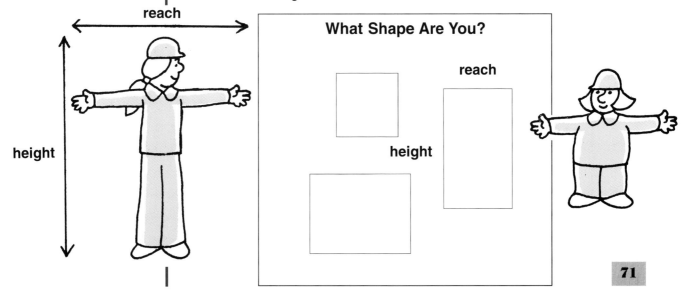

71

MEASUREMENT

ACTIVITY 2

Comparison or ratios of height to some body parts can be discovered by students working in pairs. Each student has a string equal to his or her height. Students are asked to estimate how many times a piece of string equal to their height would wrap around their head, and the estimates are recorded.

Then, students wrap the string around their heads with partners helping each other. Students will find that the string wraps about three times around the head. The comparison of these lengths can be written as ratios.

>Ratio: Head to height is 1:3 or 1/3.

Another ratio is the length of a foot to the height of the student. This ratio is usually about 1 to 6.

OBJECTIVE: Students estimate and measure lengths in centimeters.

ACTIVITY 1

Using the 30 centimeter ruler or tape, ask the students to estimate and measure the length of objects in the classroom. Have them list objects they select.

LENGTHS

	Estimate	Measurement
Pencil	_____ cm	_____ cm
Crayon	_____ cm	_____ cm
Paper clip	_____ cm	_____ cm
Math book	_____ cm	_____ cm
Audio tape	_____ cm	_____ cm
Board eraser	_____ cm	_____ cm

MEASUREMENT

ACTIVITY 2

Involve students in learning to estimate lengths in this whole class activity. Students form two columns or teams and each student has a centimeter ruler.

Hold up an object such as a book and ask the first Team A player to estimate the length of the book. For example, Team A says, "20 cm." The first Team B member now measures the length of the book, and says the book is 24 cm long. Team A is now "off" by 4 cm.

Hold up another object such as a tissue box for the same two players. The first B member gives an estimate such as 25 cm. The first A Team member measures the box and says 22 cm. Team B is "off" by 3 cm. The estimate that comes closest to the actual measurement is awarded 10 points, so Team B wins 10 points. Record the points for each team. The first team members move to the back of the line, and the next pair of students repeat the activity with you holding up other objects. The team with most points is the winner.

ACTIVITY 3

Students working in groups record estimates of strings that you hold up. Hold up strings that are 10 cm, 18 cm, 24 cm, and 12 cm in length. Each group decides together on an estimate as the individual strings are held up. The group that comes closest to the real length gets 50 points.

Next have groups draw line segments for lengths presented on cards.

| 15 cm | 20 cm | 4 cm |

MEASUREMENT

OBJECTIVE: Students use a measuring tape to measure objects that are not flat.

ACTIVITY 1

Working in pairs, students can measure parts of the body and discover a pattern.

Measure around the thumb over the bone _____ cm
Measure around the wrist over the bone _____ cm
Measure around the neck over the collar bone _____ cm
Measure around the waist _____ cm

Students can gather and record the data for each measurement from all of the members of the class to see if they find some patterns. There is a saying: "Twice around the thumb is the wrist, twice around the wrist is the neck, twice around the neck is the waist." Is this saying right for any of the students?

ACTIVITY 2

Collect a box of different sized balls – basketball, tennis ball, golf ball, volleyball, kick ball, etc.

Ask the students, working in pairs, to estimate the distance around each ball and record their guesses. Then they measure the balls, record and compare their estimates and measurements.

A collection of fruits and/or vegetables provides another interesting estimate-measure activity.

PERIMETER

The perimeter of an object is the linear distance around it. Students measure perimeter with linear units such as centimeters, meters, inches, feet or yards. For Activities 1 and 2 students will need geoboards and a supply of rubber bands.

OBJECTIVE: Students explore the meaning of the perimeter of a rectangle.

ACTIVITY 1

Using geoboards and rubber bands, students make rectangles of different sizes. Ask each student to find the perimeter of his or her rectangle by counting the spaces or segments between the nails. Then ask if there are other methods that can be used to find the perimeter. Have students share their ideas. For example, "Count one side, multiply by 2, count the adjacent side, multiply by 2. Then add." Or "Add the total spaces of each side."

MEASUREMENT

ACTIVITY 2
Using geoboards and rubber bands, ask the students to make squares and find the perimeter of their squares. First by counting the spaces or segments then by adding or multiplying.

ACTIVITY 3
Students draw different polygons (triangles, trapezoids, parallelograms, etc.) and find perimeters of these geometric shapes by measuring each side with a centimeter ruler. Remind students to label the vertices (corners) with capital letters.

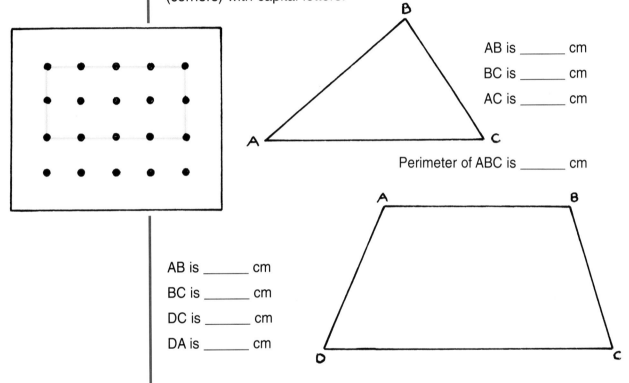

AB is _____ cm

BC is _____ cm

AC is _____ cm

Perimeter of ABC is _____ cm

AB is _____ cm

BC is _____ cm

DC is _____ cm

DA is _____ cm

Perimeter of ABCD is _____ cm

ACTIVITY 4
Make a loop of 6 feet of string. Have three students hold the string in different positions to illustrate how 6 feet in perimeter will form different kinds of triangles. Repeat this activity with four students making quadrilaterals.

MEASUREMENT

OBJECTIVE: Students show how figures with different shapes can have the same perimeter.

ACTIVITY
On graph paper ask students to draw three rectangles whose perimeter is 24 units. This requires an understanding of the relationship of length, width and perimeter in rectangles.

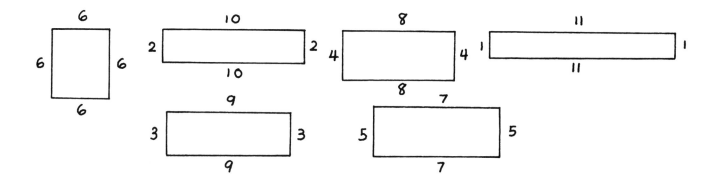

As a class activity, make a table of the values, and ask the students to find a pattern.

AREA

Area covers the surface of a geometric shape. The measurements to cover are square units.

OBJECTIVE: Students explore the meaning of area.

ACTIVITY 1
Students work in groups of four with a stack of round disks and a piece of graph paper. Ask the students to try to cover the paper with disks. Ask the students if all of the paper has been covered. Then remove the disks and ask the students to count the squares of the graph paper. Which shape is the best for measuring area? Discuss area and have students share ideas about the uses of area measurements.

MEASUREMENT

ACTIVITY 2
Using graph paper, students draw a rectangle three rows down and four units across. Have students explore the different ways to find the number of squares in the rectangle ABCD. In pairs, students write their rules for finding areas and verify these rules with different size rectangles.

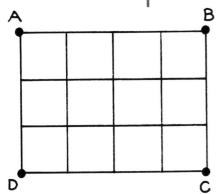

Examples:

"Add squares in each row." 4 + 4 + 4
"Multiply 3 rows of 4 squares." 3 x 4
"Add 4 sets of 3 squares." 3 + 3 + 3 + 3
"Multiply the number of squares in height by the number of squares in length." 3 x 4

ACTIVITY 3
On graph paper, students outline rectangles that have an area of 12 square units.

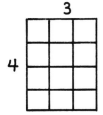
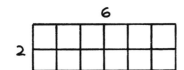
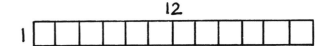

Have each student select a number of square units less than 100 and make a poster of all the possible rectangles with that area. He or she should label the length, width, perimeter and area of each.

Ask the students to write a sentence or two on their posters about what they discovered about length, width, perimeter and area.

GEOMETRY
Chapter 5

BACKGROUND

Students enjoy the study of shapes and manipulating them in a variety of ways. It is important that they learn the properties of geometric shapes and the geometric vocabulary to identify two- and three-dimensional shapes and their parts. Using this vocabulary, students can communicate with each other about the properties of geometric shapes and can describe how the shapes are alike and how they are different.

Angles and line segments are parts of most shapes. As students learn to identify right, acute and obtuse angles, they are able to classify angles and polygons. They will also become aware of parallel, intersecting and perpendicular lines and their relationships. In three-dimensional figures, students will determine the number of edges, faces and vertices.

Geometry activities should be "happenings" where students are involved in creating and investigating shapes using tangrams, geoboards, pentominoes and straws. To learn geometry, students need to investigate, experiment and explore with physical materials.

The National Council of Teachers of Mathematics has noted that students who develop a strong sense of geometry are "better prepared to learn number and measurement ideas, as well as other mathematical topics." There are also the benefits gained in terms of dealing with the "real world," which creates a convincing case for including the study of geometry in any math curriculum.

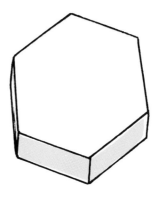

GEOMETRY

THREE-DIMENSIONAL SHAPES

Solid geometry is the study of figures that occupy space and these figures are called geometric solids. A polyhedron is a geometric solid surrounded by planes which are called *faces*. Any two faces intersect in a straight line called an *edge* and the intersection of three or more edges is called a *vertex* or *corner*.

OBJECTIVE: Students learn the meaning of the words edge, vertex and face.

ACTIVITY

Have a variety of boxes available so that students can feel the edges, vertices (corners) and faces.

For each box the edges, corners (vertices) and faces can be counted.

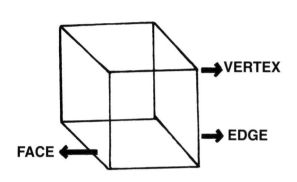

To help students remember the names, a game called "Guess My Shape?" can be played. A student picks one of the boxes and then describes it in terms of the number of edges, corners and faces it has. Another student tries to guess which shape it is from the description. Often students will use the word "side" for either edge or face, so it is important to help them communicate with the correct geometric vocabulary.

OBJECTIVE: Students identify the parts of three-dimensional figures.

ACTIVITY 1

Each student is given nine straws the same length, and pieces of pipe cleaners. Using the straws the student makes a triangular prism. Ask students to explain why they think this prism is called triangular.

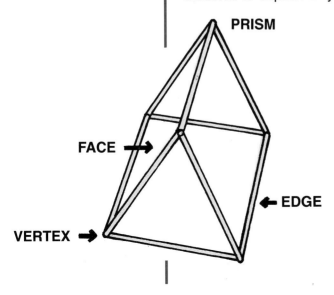

How many edges?
How many vertices?
How many faces?
What shape is each face?
What do you notice about edges in a prism?

GEOMETRY

ACTIVITY 2
Each student is given a wooden cube or die. The students count the number of edges, vertices and faces and describe the faces.

Select a variety of three-dimensional objects (boxes, books, containers, etc.) and identify the number and kinds of faces and the number of edges and vertices. Have students classify or sort them by a characteristic.

OBJECTIVE: Students create three-dimensional shapes.

ACTIVITY
Demonstrate for your students how to create three-dimensional shapes, then ask students to create their own three-dimensional shapes using toothpicks and clay. Ask the students to count the edges, corners and faces of each shape. Flexible straws also can be used.

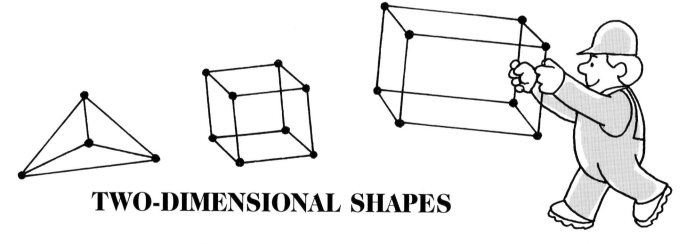

TWO-DIMENSIONAL SHAPES

Students need to learn the vocabulary of geometry to help describe the properties of two-dimensional shapes. Geometric shapes can be created in many different ways – paper strips, straws and geoboards are good for primary students. Be sure to illustrate figures in various positions, so that students become familiar with the shapes from all views.

GEOMETRY

OBJECTIVE: Students create shapes, then sort and compare these shapes.

ACTIVITY 1
Ask each student to make a three-sided shape on his or her geoboard. The students then compare their shapes and sort the geoboards by shapes that are alike and those that are different. Ask the students to describe the shapes and tell how they are alike and how they are different.

ACTIVITY 2
Using geoboards, students work in pairs. One student makes a figure and the other copies the figure so it is exactly the same in size and shape. Tell the students these figures are congruent.

OBJECTIVE: Students describe the properties of four-sided shapes.

ACTIVITY
Ask each student to make a four-sided shape on his or her geoboard. The students then compare their shapes and sort the geoboards by shapes that are alike and those that are different. Ask the students to describe the shapes and tell how they are alike and how they are different.

  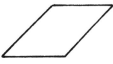  

GEOMETRY

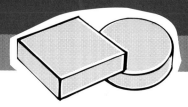

OBJECTIVE: Students learn the names of some special angles.

ACTIVITY 1
Demonstrate a right angle, then students copy the right angle using geoboards. It is important that students understand that a right angle is a corner. The geoboard can be turned so students can see that a right angle does not mean direction right.

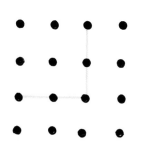
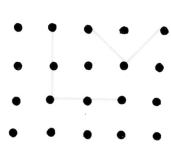

ACTIVITY 2
Using geoboards, students copy acute angles and right angles. Make several right, acute and obtuse angles in all directions and use different lengths for the sides of the angle so that students learn that the size of the angle is related to the opening.

Ask students to describe the three kinds of angles.

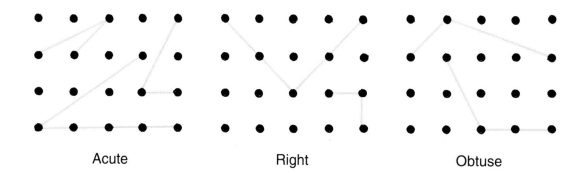

Acute   Right   Obtuse

GEOMETRY

RIGHT, ACUTE AND OBTUSE ANGLES

If the two sides of an angle are perpendicular to each other, the angle formed is a right angle. If the angle is smaller than a right angle it is called an acute angle. An angle larger than a right angle is an obtuse angle. In describing some of the properties of geometric shapes, knowing how to identify angles by their sizes as right, acute and obtuse is very important.

OBJECTIVE: Students identify right, acute and obtuse angles.

ACTIVITY 1
Make a geostrip by connecting two strips of paper 2 inches by 10 inches with a brad. Use it to make right, acute and obtuse angles.

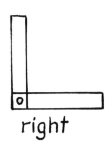
right

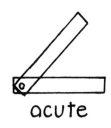
acute

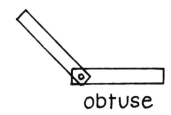
obtuse

Explain to the students that the right angle has a measure of 90 degrees. This measure describes the amount of the opening. The lengths of the sides of an angle do not affect its size or measure. Have students make angles with different length sides to verify this concept.

Have students find and identify the "square corners" on their papers, their desks or tables and in objects in the room. These are all examples of right angles.

The right angle is the reference to use in determining the kinds of other angles. An acute angle has a measure that is less than 90 degrees or is smaller than a right angle. An obtuse angle has a measure that is greater than 90 degrees or is larger than a right angle.

Be sure to show the angles in different positions so that students will recognize them in their environment.

ACTIVITY 2
Ask the students to name the angles by the hands on a clock.

3:10	_____	angle
1:25	_____	angle
5:35	_____	angle
9:00	_____	angle
2:20	_____	angle
10:05	_____	angle

GEOMETRY

 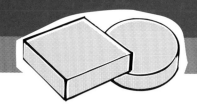

ANGLES IN GEOMETRIC SHAPES

Geometric shapes are made up of various types of angles. It is important that students can identify the various types of angles to help describe the shapes.

OBJECTIVE: Students identify various types of angles in geometric shapes.

ACTIVITY 1

Prepare a table of geometric shapes similar to the following example and ask the students to identify the number and type of angle.

SHAPES	NUMBER OF ANGLES		
	Right	Acute	Obtuse
Rectangle			
Triangle			
Parallelogram			
Pentagon			
Trapezoid			

Ask your students: "Can you make a triangle with two right angles?" Then ask students to explain why or why not.

GEOMETRY

ACTIVITY 2
Using geoboards or geopaper, students work in pairs. One student forms angles and the other identifies them by name. Next, one student states a name and the other forms the angle.

2 o'clock	_____	angle
3 o'clock	_____	angle
1:25	_____	angle
4:45	_____	angle
9 o'clock	_____	angle

ACTIVITY 3
Using a demonstration clock, show the following times on the clock and ask the students to tell whether the angle created by the hands on the clock is a right, acute or obtuse angle.

PARALLEL AND INTERSECTING LINES

If two straight lines in the same plane cross each other, they are called intersecting lines and they intersect at one point only. If two straight lines in the same plane do not intersect, they are called parallel lines.

OBJECTIVE: Students model and identify parallel and intersecting lines.

ACTIVITY 1
Ask the students to show pairs of lines on their geoboards. After comparing them, discuss what names the different pairs might have. Write "parallel" and "intersecting" on the chalkboard. Ask students to stand with their pairs next to the label.

Discuss the various places in everyday life where students see parallel and intersecting lines.

ACTIVITY 2
Ask students to use three or more straws or pencils to show parallel and intersecting lines in all directions.

85

GEOMETRY

 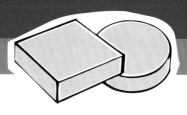

INTERSECTING, PARALLEL AND PERPENDICULAR LINES

Geometric shapes are made up of angles and lines. Two or more lines can be intersecting, parallel or perpendicular. Perpendicular lines are intersecting lines that meet to form right angles. Make certain that students form intersecting lines that form other size angles.

OBJECTIVE: Students draw, make and identify intersecting, parallel and perpendicular lines.

ACTIVITY 1

Demonstrate and identify for your students examples of intersecting, parallel and perpendicular lines. Then ask the students, using geoboards or graph paper, to make intersecting, parallel and perpendicular lines. Continue this activity by asking:

- "What type of angles can be formed by intersecting lines?"
- "What type of angles can be formed by perpendicular lines?"

As an extended activity, find items in the room that model parallel lines, intersecting lines and perpendicular lines.

ACTIVITY 2

In this activity, each pair of students is given 12 toothpicks. Player A tosses the 12 toothpicks on a flat surface. Then Player A identifies any parallel, intersecting or perpendicular lines. Five points are awarded for parallel lines, one point for intersecting lines that are not perpendicular and ten points for perpendicular lines. The player records his or her total points.

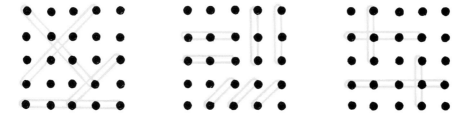

(1)   (5)   (10)   (1)   (5)   (1)

Then Player B collects the toothpicks and proceeds in the same manner as Player A. The player with the most points wins the round. Five rounds can be played to determine the grand winner.

GEOMETRY

ACTIVITY 3

In this activity, students again work in pairs. One student draws a shape such as a quadrilateral (four sides). The other student cannot see the drawing. The student who drew the shape describes the figure in terms of lines and angles. The other student draws a figure according to the description.

Description:
One pair of parallel lines.
Four sides.
Two acute angles on the bottom and two obtuse angles on the top.

LINE SYMMETRY

A line of symmetry separates a figure into two matching or congruent shapes. The part on one side of the line is the exact opposite in direction of the part on the other side of the line.

OBJECTIVE: Students make lines of symmetry by folding activities.

ACTIVITY

Each student will need two pieces of paper and scissors. Tell the students to fold one piece of paper in half horizontally and cut a shape on the fold. Students open up the shape. The fold is a line of symmetry.

Fold

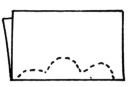
Cut along the fold

Line of symmetry

Tell the students to fold the second piece of paper in half vertically, cut a shape on the fold and open the shape to see the line of symmetry.

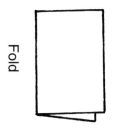
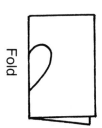
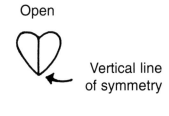

Open

Vertical line of symmetry

GEOMETRY

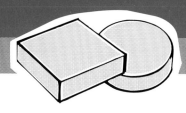

Note the direction of the fold does not affect the symmetry of the figure. Rotate the shapes so that students do not expect to see the shapes and lines of symmetry in a fixed position.

Ask the students to fold a piece of paper twice and then cut a shape starting on one fold and ending on the other fold. This will give two lines of symmetry on the opened figure.

OBJECTIVE: Students draw lines of symmetry on capital letters and pictures.

ACTIVITY 1

Provide a paper with large printed capital letters. The students determine whether lines of symmetry can be drawn on each letter. Folding on the drawn line will verify their work.

Which letters have one line of symmetry, two lines or no lines of symmetry?

A	B	C	D	E
K	H	M	Y	S
N	V	W	X	O	Z

ACTIVITY 2

Ask students to collect pictures for a symmetry booklet. On each picture students draw the lines of symmetry.

FRACTIONS
Chapter 6

BACKGROUND

Students must develop an understanding of the meaning of fractional numbers and the symbols for them. Often students have difficulty in later grades with the operations of fractions because they lack the basic understanding of fractions.

The physical modeling of fractions is necessary since the meaning of these numbers does not come from the usual sequential counting. The common model is the region outlined by geometric shapes. Other models used regularly are sets of objects and the number line. All three concrete situations are based on separating into fair shares or equal parts and naming those parts.

Rulers, gauges, measuring cups and beakers are all everyday tools that use fractions. Provide opportunities for students to measure fractional amounts with these tools.

Comparing and ordering fractions and naming equivalent fractions are important concepts that are also learned with physical models. Equivalent fractions are needed in computing with fractions.

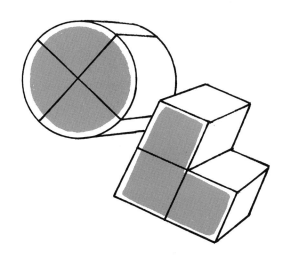

FRACTIONS

MEANING OF FRACTIONS

Fraction concepts can be introduced and clarified through activities that demonstrate fair shares. Students are familiar with the idea of equal parts or fair shares.

OBJECTIVE: Students recognize fair shares of a whole.

ACTIVITY
Prepare a variety of geometric shapes and cut some of them into equal parts and others into unequal parts. Ask the students to identify those shapes that have been cut into fair shares.

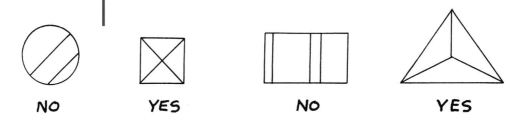

NO   YES   NO   YES

OBJECTIVE: Students record the number of fair shares.

ACTIVITY 1
Prepare a worksheet of geometric shapes similar to the example below. Ask the students to count the number of fair shares in each shape and write that number in the box below the line. Encourage the students to use the ordinal number names by asking questions such as: "How many eighths are there?" "How many thirds?"

90

FRACTIONS

NAMING FRACTIONS

Explain to the students that fractions are made up of two numbers, one written above a horizontal bar and the other written under it. The denominator which is the number below the bar indicates the number of equal parts or fair shares in a region or set.

The six equal parts are each called a "sixth."

The two equal parts are called "halves."

There are five objects in the circle. Each is called a "fifth."

As students record the number of fair shares, they also learn that the numbers in the denominator are read differently. Except for 2, the denominators are read as ordinal numbers – third, fourth, fifth, sixth, etc.

The number above the bar is the numerator. It indicates the number of special parts in the region or set.

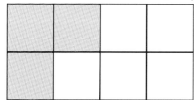
3 of 8 squares are shaded   3-eighths

4 out of the 6 objects are stars   4-sixths

The name of each equal part is called a unit fraction. A unit fraction always has one as the numerator. Demonstrate this by showing the students a circle divided into fourths as illustrated and explain that each part is one-fourth of the whole circle.

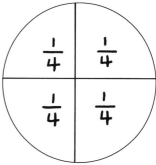

When the numerator is zero, the fraction represents zero and when the numerator is the same as the denominator, the fraction represents one.

91

FRACTIONS

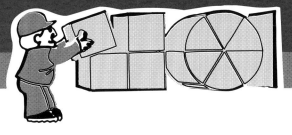

ACTIVITY
After the students are confident in using ordinal number words, state the name of a fair share, such as thirds, fourths, etc., and ask the students to create that fair share using the parts of equal geometric shapes.

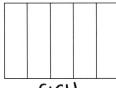 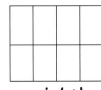

    fifths   eighths   fourths

OBJECTIVE: Students write the fraction shown by a geometric shape.

ACTIVITY
Prepare students' worksheets similar to the sample below. Ask the students to write the number of fair shares next to each shape. Then count the number of shaded fair shares and write that number in the space above it. Ask students to write the fraction name of the denominator. This distinguishes between the cardinal number that tells how many parts and the ordinal that names the kind of parts.

  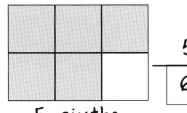

                  3-fourths   5-sixths

OBJECTIVE: Students express 1 as a fraction in different ways.

ACTIVITY
Prepare a variety of geometric shapes divided into shaded fair shares on one side with the number 1 written on the reverse side as shown below. Ask the students to write the number of the fair shares and the number of shaded shares and then turn the card over and write the number 1.

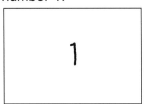 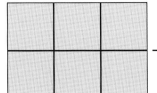

Ask students to fold a paper once and write the fraction name for the paper. $\frac{2}{2} = 1$

Folding and labeling again and again will enable students to record many names for 1.

$$\frac{2}{2}, \frac{4}{4}, \frac{8}{8}, \frac{16}{16}$$

FRACTIONS

OBJECTIVE: Students write fractions to match a model of the fraction.

ACTIVITY
Divide the students into two teams. Show each team a geometric shape with shaded parts. The students take turns writing the fractions shown on the models. A team gets ten points for each correct fraction and loses ten points for each incorrect fraction.

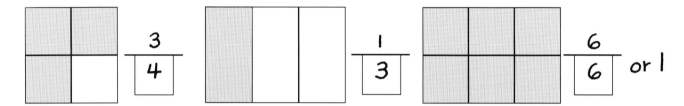

OBJECTIVE: Students compare fraction models and determine whether they are greater or less than $\frac{1}{2}$.

ACTIVITY
Demonstrate the fraction one-half by showing a geometric shape with one-half shaded. Show the students same sized shapes and ask the students to say whether the shaded area is greater or less than one-half of that unit.

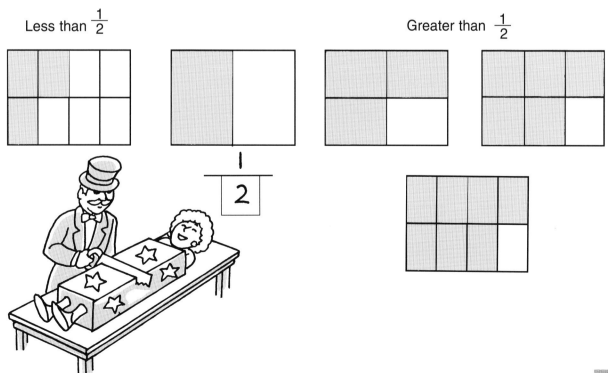

93

FRACTIONS

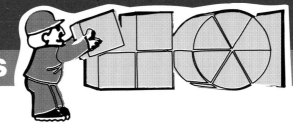

OBJECTIVE: Students identify numerator and denominator and read or write fractional numbers.

ACTIVITY 1
Show the students a geometric shape divided into four parts.

Ask, "What part of the square is shaded?" Since there are four parts and none of the parts are shaded, the answer is $\frac{0}{4} = 0$.

Then ask: "What part of the rectangle is shaded?" Since there are four equal parts and four parts are shaded, the answer is $\frac{4}{4}$. This represents all of the figure or 1.

ACTIVITY 2
Ask the students to mark off equal parts on rows or on geometric shapes and then write the word name for the number and the kinds of parts as shown in examples 1 and 2.

Example 1

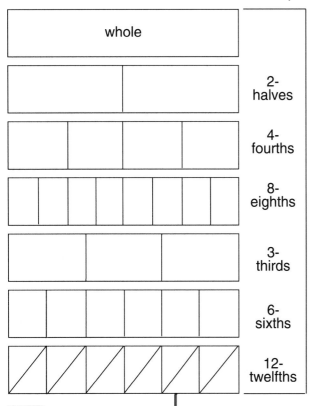

Example 2

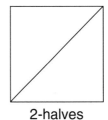

2-halves   4-fourths   16-sixteenths

94

FRACTIONS

ACTIVITY 3
Use the entire class, special rows or designated tables as a unit or set. Ask students from that group who fit a certain description or characteristic to stand. For example, ask students to stand who:

- have brown eyes.
- have a bird as a pet.
- had cereal for breakfast.
- have buttons on their clothes.
- have three different vowels in their name.

Then ask the students what fractional part these students are of the larger group.

ASSESSING THE UNDERSTANDING OF FRACTIONS

It is important to periodically assess each student's understanding of basic fraction concepts before continuing with more advanced concepts. Many of the difficulties students experience with fractions can be traced to an inadequate understanding of these basic concepts.

OBJECTIVE: Students draw a picture or make a model for a fraction demonstrating their understanding of the meaning of fractions.

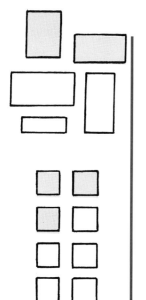

ACTIVITY 1
Ask students to draw pictures to show the fractional parts named.

Examples:

- $\frac{2}{5}$ of a set of rectangles are shaded
- $\frac{3}{8}$ of the squares are red
- $\frac{5}{9}$ of the fruits are apples
- $\frac{1}{2}$ of the square is blue
- $\frac{2}{3}$ of the circle is shaded

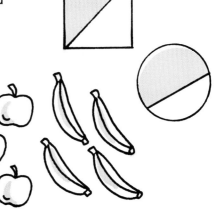

FRACTIONS

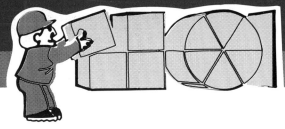

ACTIVITY 2
Ask the students to tell how the fractions illustrated below are different and how they are alike.

$\frac{0}{3}$ and $\frac{0}{4}$ $\frac{6}{6}$ and $\frac{8}{8}$

Zero-thirds and zero-fourths show that something is divided in thirds and something else divided in fourths. They are alike because both represent the number zero. Six-sixths and eight-eighths show that something is divided in sixths and something is divided into eighths. They are alike because both fractions represent the number one.

OBJECTIVE: Students use letters to build and order fractions in a group activity.

ACTIVITY
Each student counts the number of letters in his or her last name. Students with the same number of letters will form groups. This number will be the denominator of their fractions.

The number of consonants in each student's last name will be the numerator of his or her fraction.

$$\frac{\text{number of consonants}}{\text{number of letters}}$$

In each group, students compare their fractions and order them from least to greatest. Each group explains the results to the other groups.

VARIATIONS
- Use the number of vowels as the numerator.
- Use the letters in the first name.
- Identify all fractions that are names for one-half.

"LOLA", $\frac{2}{4}$ OF THE LETTERS ARE VOWELS, $\frac{2}{4}$ IS A NAME FOR $\frac{1}{2}$.

"LAUREL" $\frac{3}{6}$ OF THE LETTERS ARE VOWELS. $\frac{3}{6}$ IS A NAME FOR $\frac{1}{2}$.

Discuss why $\frac{1}{2}$ can represent different values.

EQUIVALENT FRACTIONS

The following activities are designed to introduce equivalent fractions using concrete models. In these activities, students discover and demonstrate that certain fractions have equal value.

OBJECTIVE: Students develop an understanding of equivalent fractions.

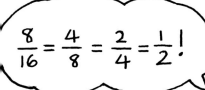

ACTIVITY 1

Each student will need a sheet of one-inch graph paper and scissors.

- Mark a strip that is 16 squares long. Write the number 1 in the middle of the strip.

- Mark a strip that is 16 squares long and then mark the strip after 8 squares. Write $\frac{1}{2}$ on each strip.

- Mark a strip 16 squares long. After 4 squares make a mark so there are 4 pieces 4 squares long. Write $\frac{1}{4}$ on each piece.

- Mark a strip 16 squares long. After 2 squares make a mark so there are 8 pieces 2 squares long. Write $\frac{1}{8}$ on each piece.

- Mark a strip 16 square long. Mark into 16 pieces 1 square long. Write $\frac{1}{16}$ on each piece.

- Cut the rows of 16 squares and then cut into the fraction parts.

1															
$\frac{1}{2}$								$\frac{1}{2}$							
$\frac{1}{4}$				$\frac{1}{4}$				$\frac{1}{4}$				$\frac{1}{4}$			
$\frac{1}{8}$		$\frac{1}{8}$		$\frac{1}{8}$		$\frac{1}{8}$		$\frac{1}{8}$		$\frac{1}{8}$		$\frac{1}{8}$		$\frac{1}{8}$	
$\frac{1}{16}$	$\frac{1}{16}$	$\frac{1}{16}$	$\frac{1}{16}$	$\frac{1}{16}$	$\frac{1}{16}$	$\frac{1}{16}$	$\frac{1}{16}$	$\frac{1}{16}$	$\frac{1}{16}$	$\frac{1}{16}$	$\frac{1}{16}$	$\frac{1}{16}$	$\frac{1}{16}$	$\frac{1}{16}$	$\frac{1}{16}$

FRACTIONS

Using the fraction strips, the students have models to help find and understand equivalent fractions. The pieces can be used to match the pieces named in an exercise.

$\dfrac{1}{2} = \dfrac{?}{8}$

$\dfrac{1}{2}$			
$\dfrac{1}{8}$	$\dfrac{1}{8}$	$\dfrac{1}{8}$	$\dfrac{1}{8}$

$\dfrac{1}{2} = \dfrac{4}{8}$

1-half equals 4-eighths

The students can store the strips in an envelope to use in other fraction activities.

ACTIVITY 2

Remind the students that the denominator names the kind of parts and the numerator tells how many. Then ask the students to find the equivalent fraction using their fraction strips. Ask the students to identify patterns in equivalent fractions.

$\dfrac{1}{2} = \dfrac{?}{4}$ $\dfrac{1}{2} = \dfrac{?}{8}$ $\dfrac{1}{4} = \dfrac{?}{8}$

$\dfrac{3}{6} = \dfrac{4}{?}$ $\dfrac{3}{?} = \dfrac{8}{16}$ $\dfrac{12}{16} = \dfrac{?}{8}$

ACTIVITY 3

Ask students to find other names for $\dfrac{1}{2}$.

$\dfrac{50}{?}$ $\dfrac{2}{4}$ $\dfrac{?}{20}$

$\boxed{\dfrac{1}{2}}$

$\dfrac{4}{8}$ $\dfrac{12}{?}$ $\dfrac{5}{?}$

$\dfrac{?}{100}$ $\dfrac{12}{?}$ $\dfrac{9}{?}$

$\dfrac{3}{6}, \dfrac{6}{12}, \dfrac{7}{14}, \dfrac{9}{18}, \dfrac{10}{20}, \dfrac{11}{22}, \dfrac{12}{24}, \dfrac{13}{26}, \dfrac{14}{28} \cdots \cdots$ ALL EQUAL $\dfrac{1}{2}$!

FRACTIONS

OBJECTIVE: Students identify equivalent fractions using number strips.

ACTIVITY

Each student will need a sheet of one-inch graph paper. Ask students to write the product of the multiplication tables two through nine on rows of the graph paper.

2	4	6	8	10	12	14	16	18	20
3	6	9	12	15	18	21	24	27	30
4	8	12	16	20	24	28	32	36	40
5	10	15	20	25	30	35	40	45	50
6	12	18	24	30	36	42	48	54	60
7	14	21	28	35	42	49	56	63	70
8	16	24	32	40	48	56	64	72	80
9	18	27	36	45	54	63	72	81	90

Students cut the rows apart and use the strips to find the equivalent fractions of a given fraction. Ask the students to place the three strip above the eight strip.

3	6	9	12	15	18	21	24	27	30

8	16	24	32	40	48	56	64	72	80

Ask the students to compare the fractions formed using the strips.

$$\frac{3}{8} = \frac{21}{?} \qquad \frac{3}{8} = \frac{?}{40}$$

After students have found that the pairs are equivalent fractions, they should tell which numbers could be used to multiply both the numerator and denominator of $\frac{3}{8}$ to get the other fractions.

$$\frac{3}{8} \times \frac{7}{7} = \frac{21}{56}$$

FRACTIONS

COMPARING AND ORDERING FRACTIONS

It is helpful to give students as many opportunities as possible to compare and order fractions to promote their understanding of fractions. The following activities can also be used to assess their understanding of fractions.

OBJECTIVE: Students compare and order fractions.

ACTIVITY

Using the fraction strips, students decide which fractions are equal (=), and which fraction is less than (<) or is greater than (>) the other. Remind the students that only fractions that represent parts of the same unit can be compared. Prepare a worksheet similar to the example given below.

Write >, < or = to make each fraction sentence true.

$\frac{1}{4}\bigcirc\frac{4}{16}$ $\frac{1}{2}\bigcirc\frac{5}{8}$ $\frac{3}{8}\bigcirc\frac{1}{4}$

$\frac{7}{16}\bigcirc\frac{3}{4}$ $\frac{4}{16}\bigcirc\frac{2}{4}$ $\frac{5}{16}\bigcirc\frac{1}{2}$

OBJECTIVE: Students sort fractions by size into sets.

This group activity can be used to assess students' understanding of the order of fractions.

ACTIVITY

Provide each group of three or four students with 15 fractions written on cards. All fractions should be less than one.

In each group, students take turns placing a fraction card into the appropriate set and explaining the selection.

Close to 0	Close to $\frac{1}{2}$	Close to 1
$\boxed{\frac{1}{4}}$	$\boxed{\frac{3}{8}}$	$\boxed{\frac{7}{8}}$

After the group agrees on the choices, the students record what they did and then describe how they sorted the fractions.

FRACTIONS

OBJECTIVE: Students demonstrate their understanding of fractions by combining fractional parts of two words to create a new word.

ACTIVITY

Tell the students to count the number of letters in the two words, then find the fraction of each word and write the new word on the line.

Example: The first $\frac{1}{3}$ of **bat** combined with the last $\frac{2}{3}$ of **toy**.

The word is **boy**.

Find the word:
1. The first $\frac{2}{3}$ of **ten** combined with the last $\frac{2}{3}$ of **Sam**. _____
2. The first $\frac{2}{5}$ of **skill** combined with the last $\frac{1}{2}$ of **drip**. _____
3. The first $\frac{1}{2}$ of **blue** combined with the last $\frac{3}{4}$ of **send**. _____
4. The first $\frac{1}{2}$ of **take** combined with the last $\frac{3}{5}$ of **fable**. _____
5. The first $\frac{1}{2}$ of **street** combined with the last $\frac{3}{4}$ of **ripe**. _____
6. The first $\frac{1}{3}$ of **car** combined with the last $\frac{3}{4}$ of **told**. _____
7. The first $\frac{3}{5}$ of **speed** combined with the last $\frac{2}{3}$ of **ant**. _____
8. The first $\frac{2}{5}$ of **would** combined with the last $\frac{2}{3}$ of **rod**. _____
9. The first $\frac{3}{5}$ of **price** combined with the last $\frac{1}{2}$ of **rant**. _____
10. The first $\frac{2}{6}$ of **friend** combined with the last $\frac{4}{5}$ of **night**. _____
11. The first $\frac{1}{2}$ of **travel** combined with the last $\frac{1}{2}$ of **rain**. _____
12. The first $\frac{2}{5}$ of **tools** combined with the last $\frac{1}{3}$ of **boy**. _____
13. The first $\frac{2}{7}$ of **special** combined with the last $\frac{3}{6}$ of **friend**. _____
14. The first $\frac{1}{8}$ of **together** combined with the last $\frac{3}{4}$ of **ball**. _____

Make up your own "fraction words."

IMPROPER FRACTIONS/MIXED NUMBERS

Proper fractions are fractions in which the numerator is less than the denominator. A proper fraction has a value less than 1. An improper fraction is a fraction in which the numerator is equal to or greater than the denominator. An improper fraction has a value equal to or greater than 1. Improper fractions are often expressed as mixed numbers.

OBJECTIVE: Students use models for naming improper fractions or mixed numbers.

FRACTIONS

ACTIVITY 1

Give each student two pieces of square paper. Ask the students to write the number one on one side of each sheet. Then the students fold each paper into fourths.

On one sheet, the student shades the four fourths and on the other sheet he or she shades three fourths. Then the student writes the improper fraction seven fourths or $\frac{7}{4}$.

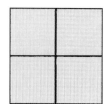   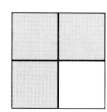

The student then turns over the paper with the four shaded fourths and sees the number 1. The student then writes the mixed number 1 3/4. Explain to the students that an improper fraction is equal to or greater than one and the denominator is less than the numerator. A mixed number is greater than one and is written with a whole number and a fraction.

   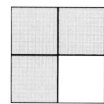 $\frac{7}{4} = 1\frac{3}{4}$

Have students fold and label other sheets to model improper fractions.

ACTIVITY 2

Using their fractional strips, two students work together to show improper fractions and then rename these fractions as mixed numbers.

$\frac{9}{8}$ = _____ _____   Lay out 9 eighths. Place the 1 strip on top to demonstrate 1 and $\frac{1}{8}$.
Write the mixed number.

$\frac{7}{4}$ = _____ _____   Lay out 7 fourths. Place the 1 strip on top to demonstrate 1 and $\frac{3}{4}$.
Write the mixed number.

$\frac{3}{2}$ = _____ _____

$\frac{17}{16}$ = _____ _____

FOURTHS?
SIXTHS?
TENTHS?
EIGHTHS!

PROBLEM SOLVING

Chapter 7

BACKGROUND

A mathematical problem is defined as a question or situation where an available solution is not known immediately. The problem solver is seeking a goal or a solution that requires skills such as:

- decision in the trial and improve strategy
- making predictions from data
- evaluating the reasonableness of the solution.

In order to function in a changing society, people need to solve a variety of problems. Students need to become effective and confident problem solvers. Whether or not a situation is a problem is an individual matter depending on the individual's reaction, understanding of the situation or available skills. Some problems can be real-life problems that appear in the daily lives of the students.

Other problems can be contrived problems that will help to build problem-solving abilities. Others are applications of mathematical concepts and operations.

Students need many opportunities to solve all types of problems. Various types of games can be used as problem-solving situations. To win the games students need to analyze the plays, predict the results and try various strategies. In school work, word problems are only one part of the problem-solving process. A major objective of word problems is to have the student decide on the strategy needed to solve the problem. A secondary objective is the arithmetic operation the student must perform to solve the problem. Other objectives are to label the result correctly and to determine if the answer is reasonable.

PROBLEM SOLVING

PATTERNS IN PROBLEM SOLVING

Mathematics is made up of patterns of numbers and patterns in geometry. Patterns weave mathematical ideas together.

Working with patterns nurtures the type of thinking often used in solving problems. You can help students solve a problem by suggesting: "Make a table with the data and look for a pattern." Patterns often become the key to the solution of a problem.

OBJECTIVE: Students identify patterns by observing and copying actions.

ACTIVITY 1

In action patterns students follow given patterns which they see or hear and then describe the patterns in words. Patterns are repeated many times. Students can create simple action problems after some practice.

Some ideas for patterns:

- clap twice, touch your head
- snap, snap, snap, clap, clap
- clap hands, tap feet, hand on hip
- use a song such as "The Hokey Pokey" with motions that create a pattern

ACTIVITY 2

Picture cards of actions help students establish patterns that are not teacher directed.

Students in small groups select two or three of the cards that show actions, then repeat with their actions in a sequence.

PROBLEM SOLVING

PATTERNS WITH OBJECTS

There are many types of objects that can be used for patterning activities such as blocks, rods, buttons, geometric shapes, small toys, shells and toothpicks. The type of material is not important since the development of the ability to recognize, describe and continue patterns is the objective.

OBJECTIVE: Students recognize, describe and repeat a given pattern.

ACTIVITY
Students are given a pattern that is repeated at least twice and then describe and repeat the pattern.

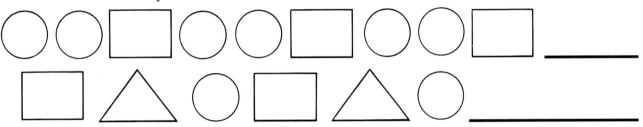

In pairs or groups, students can make patterns for other students to recognize and repeat.

OBJECTIVE: Students create a pattern with different materials that is similar to a given pattern.

ACTIVITY
Working with another student, each student creates a pattern using materials. The second student creates a similar pattern using a different material. The students then tell how their patterns are alike and how they are different.

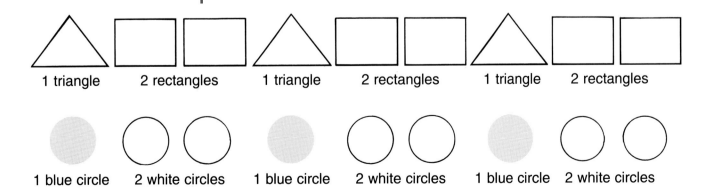

105

PROBLEM SOLVING

OBJECTIVE: Students complete a pattern that has been repeated at least twice.

ACTIVITY
Prepare a pattern such as the example below. Then ask the students to describe the beginning and the end of the pattern and then complete the pattern that has been started.

OBJECTIVE: Students find and describe patterns in their environment.

ACTIVITY
- Ask students to describe to the class the patterns they find.
- Designs in floor covering, wallpaper, clothing, art work, buildings, grocery stores, etc., will provide an unending supply of patterns.
- Make a classroom collage or display of samples of patterns that the students have found.
- Students can make their own booklets of patterns as a project.

ORGANIZING DATA

An important part of the problem-solving process is gathering data and making decisions based on the data. In the primary grades data are gathered on many different topics and then the data can be recorded in a graph. Students can answer questions and ask questions about the data shown in the graph.

The bar graph is used in the primary grades to record data. Students learn that each graph needs a base line, a title and type of data. Venn diagrams are also pictorial ways of organizing information. Students learn to record data in proper parts of the diagram.

OBJECTIVE: Students gather data and record the data in a graph.

PROBLEM SOLVING

Color of Hair

ACTIVITY 1

Give each student a large sheet of construction paper, and ask each student to draw and color his or her portrait and print his or her name on the paper. Give directions for position of the paper so that the pictures are uniformly aligned either vertical or horizontal. Tell the students they are going to sort their portraits by the color of their hair. On the floor, make a base line and label the choices.

Each student places his or her picture on the appropriate column. Discuss the number in each column. Compare columns, find how many more or less in each column.

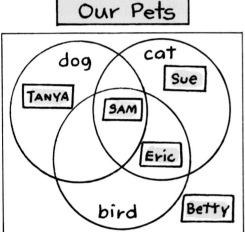

Use this type of exercise daily as a way of organizing information.
Other categories:

- favorite vegetable
- my pet
- kind of shoes

Ask your students to select other topics.

Data also can be organized and recorded in a Venn Diagram. Each student can place a Post-It Note with his or her name on the appropriate location. Each student must decide on the location for his or her name. Pose questions about the data shown. Have students explain what they learn from the data.

ACTIVITY 2

Unifix Cubes or blocks can be used to make towers to demonstrate a bar graph.

Prepare Favorite Fruit labels and ask students to place Unifix Cubes in the columns of their favorite fruits.

After the bar graph has been completed, analyze the graph by playing the game, "What is the Question?"

- "The answer is four, what is the question?"
  (How many students like bananas?)

- "The answer is 16, what is the question?"
  (How many students prefer apples or oranges?)

- "The answer is two what is the question?"
  (How many more students like apples than oranges?)

- "The answer is 22. What is the question?"

107

PROBLEM SOLVING

ACTIVITY 3

Using one-inch graph paper, prepare a Favorite Color sheet as shown. Ask your students to vote for their favorite color, and complete the graph with crayon or marker. Then ask questions such as:

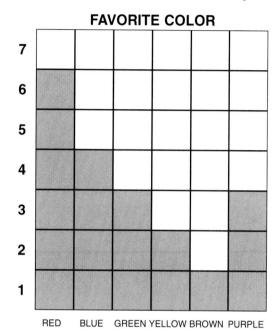

"The answer is three, what is the question?" (How many students voted for green or purple?) or (How many students voted for brown and yellow?)

Often there is more than one question for the given answer. Students can state answers and other students ask the questions.

OBJECTIVE: Students record information in Venn Diagrams and describe the data collected.

ACTIVITY

Using yarn, make two large circles that overlap on the floor and make category cards that say, for example, ice cream and cake. Ask 12 students to stand in the Venn Diagram showing their favorite dessert. Then ask:

- "How many students like ice cream?" (8)
- "How many like cake?" (7)
- "Does 8 + 7 equal 12?"
- "Where did the three extra children come from?"

This activity can be repeated using three overlapping circles and three selection topics.

PROBLEM SOLVING

WORD PROBLEMS

The question is an important part of a word problem. Students need to be involved in writing questions and also in posing word problems in which they have an interest.

OBJECTIVE: Students practice posing questions for word problems.

ACTIVITIES
- Give students an answer such as: "The answer is 36." Ask the students to either write or orally state questions that have an answer of 36. All of the students should be involved in this activity.
- Give the students an incomplete word problem. The students pose questions that pertain to the unfinished word problem. Many different questions will complete the problem.

Example:
There are twenty-five students in the class. Each student contributes $1.50 for a class party.

OBJECTIVE: Students write word problems.

SMALL GROUP ACTIVITY
- Give the students an equation and ask them to write a word problem with a question that would use the equation for finding the solution.

    Example:
    9 + 6 = 15

The students then read and discuss their stories within the small groups. The students can also make up an equation and write stories to go with the equation.

109

PROBLEM SOLVING

CLASS ACTIVITY

Ask each student to bring in a newspaper ad that has an item for sale. Students should paste the ad on a sheet of paper and write a word problem that uses the information in the ad. Students then write the solution to the word problem on the back of the paper. Each page is titled by the "author" and the students can exchange their stories and find the solutions. The pages can become a problem-solving file for the class.

Example: 3 cartons cost $1.00
How many cartons can you buy for $7.50?

NUMBER PUZZLES

Number puzzles offer students the opportunity to practice the strategies that prove helpful in problem-solving situations. These simple puzzles can be used as class activities and the students can then compare the strategies they used.

OBJECTIVE: Students use trial and improve strategies to find solutions to number puzzles.

ACTIVITY 1

Each student will need a supply of chips or paper disks with the numbers 1, 2, 3, 4, 5 and 6 on them. Prepare and distribute to the students a number puzzle with circles large enough to accommodate the chips or disks.

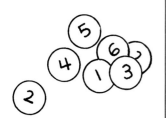

Ask the students to place numbers in the pattern of circles to make the sum of 9 in all three directions.

Now try:
Make a sum of 10 in all three directions.
Make a sum of 11 in all three directions.
Make a sum of 12 in all three directions.

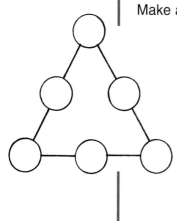

- Ask the students if they can make the sum of 13 in all directions. Why? Why not?
- Using the numbers 2, 3, 6, 8, 10, 12, what sums can be made on this pattern?

110

PROBLEM SOLVING

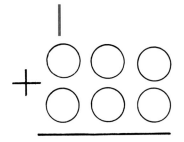

ACTIVITY 2 – Nine Circle Puzzle
In this activity students will use chips or paper disks with the numbers 1, 2, 3, 4, 5, 6, 7, 8 and 9 and a puzzle sheet with nine circles large enough to accommodate the chips or disks.

Place numbers in the circles to make an addition problem that requires renaming once.

Find 10 solutions.

Tell your students that there are over 300 solutions and 32 different sums. How many can you find in a week?

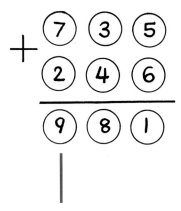

SMALL GROUP PROBLEM SOLVING

OBJECTIVE: Students find different ways to find the sum of a set of given numbers.

ACTIVITY
Given the numbers 1 + 2 + 3 + 4 + 5 + 6 + 7 + 8 + 9, ask the students to look for patterns and discuss different solutions. For example, a student might put all the tens together first (1 + 9, 2 + 8, 3 + 7, 4 + 6), then add the tens and the remaining number 5.

The question could then be asked: "Which of the solutions you found for the set of numbers 1 to 9 would give a simple solution to add all the numbers from 1 to 100?"

111

PROBLEM SOLVING

GAMES

Research has shown that children who have been involved in playing strategy games have improved their problem-solving ability. Games have a strong appeal for people of all ages. An entirely new wave of strategy games using the personal computer is now appearing.

To students, games are real-world problem situations. They want to win and they enjoy playing games. Traditional games of dominoes, Monopoly and dice games are excellent for family and class activities.

Students should learn to analyze and discuss their plays and learn to record their moves in each game. Students need to play each game several times to refine their strategies. They learn from each other as they discuss with the class the strategies that consistently result in winning.

OBJECTIVE: Students discover and practice strategies to win a game.

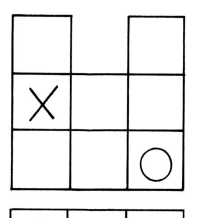

ACTIVITY 1 – Valley Tic-Tac-Toe

Valley Tic-Tac-Toe is a variation of the game tic-tac-toe. Most students know how to play tic-tac-toe and this variation of the game should create an interest in playing strategies. By eliminating one square of the tic-tac-toe grid, the game becomes more difficult since there are two less paths which reduce the ways a player can win.

Two players take turns placing their colored markers "x" and "o" in the squares. The winner is the first person to get three of his or her markers in a row horizontally, vertically or diagonally. There are six possible ways to win.

ACTIVITY 2 – Jest

Jest is a blocking strategy game for two players. The game is played on a 3 x 3 array of squares.

Each of the two players has three chips or markers of a single color. The starting position is as shown. The players take turns moving one of their markers. Each marker may be moved to another square in any direction — forward, backward, horizontally or diagonally. Markers cannot jump over other markers nor can any markers be captured. The winner is the first player to have all three of his or her markers in the opponent's starting line.

PROBLEM SOLVING

ACTIVITY 3 – Nim

Nim is a capture strategy game that is played with a set of eleven chips, bottle caps or other markers.

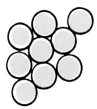

The chips or markers are placed on a table between two players. In turn, each player chooses to pick up one, two or three chips. The winner is the player who picks up the last chip. The game may also be played so that the player who picks up the final chip is the loser.

NUMBER SENSE
Chapter 8

BACKGROUND

The best way to develop number sense in students is to start them thinking about numbers. Students working alone or in small groups, can be involved in activities that require observation, comparison and uncovering assumptions. Students who have developed number sense know that they can control numbers.

Number sense is understanding the concepts in measurement, in data, in geometry, as well as in computation. Students who have number sense have developed flexible ways to work with numbers that go beyond depending on standard algorithms.

Estimation skills with both number and measurement are important. In the real world when these skills are applied it is often a more appropriate solution to a problem. Students need to learn that exactness is not the only part of solutions in life situations.

Geometry and measurement are regularly used in real life. The vocabulary used in geometry is needed to describe the properties of geometric shapes. Measurement of geometric shapes must develop from measuring with non-standard units to learning the formulas of perimeter, area and volume.

NUMBER SENSE

ODD AND EVEN NUMBERS

It is important that students achieve a sense of competence in identifying patterns in adding even and odd numbers. The following activities illustrate the kinds of sums that result from adding the different combinations of odd and even numbers.

OBJECTIVE: Students use models to find patterns for adding odd and even numbers.

ACTIVITY 1

Ask your students to cut out shapes of two rows or two columns for the numbers 1 to 8 from a piece of one-inch graph paper as illustrated.

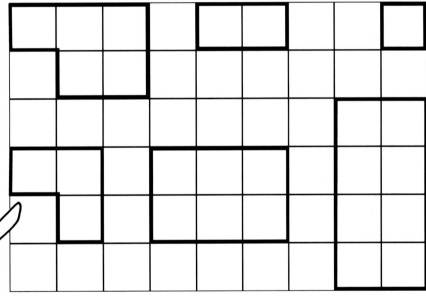

Ask the students to sort the shapes into two sets – one set of even shapes and the other set of odd shapes. How do all the odd shapes look? Then the students should pick two shapes – one even shape and one odd shape – and connect the two shapes.

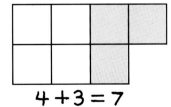

4 + 3 = 7

Ask the students:

- "What type of shape is the result?"
- "The sum of an even number and an odd number is what kind of number?"
- "Test your answer with larger odd and even numbers."

115

NUMBER SENSE

Next, ask the students to connect any two odd shapes. "What type of shape does it become?"

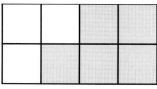 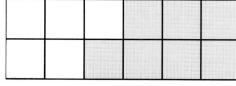

$3+5=8$ $5+7=12$

The sum of two odd numbers is an even number. Have students test this with pairs of larger numbers.

Then students should connect any two even shapes.

"What type of shape does it become?"

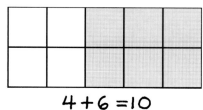 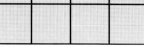

$4+6=10$ $2+8=10$

The sum of two even numbers is an even number. Again, test this with many pairs of larger numbers.

ACTIVITY 2

Prepare a table to illustrate the preceding conclusions for the addition of odd and even numbers. Ask the students to complete the table by writing in E for even or O for odd.

Students fill in:

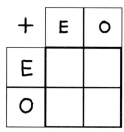 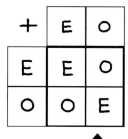

↑
Odd + Odd = Even

NUMBER SENSE

NUMBER GAMES

While using different strategies to play number games, students gain a better understanding of working with number operations. The following activities offer students opportunities to use all of the mathematical operations in entertaining and challenging game activities.

OBJECTIVE: Students use a variety of strategies to play number games.

ACTIVITY 1 – Krypto

Introduce the game of Krypto.

Have students choose five numbers from 1 through 20. After the five numbers are chosen, another student chooses one number that is between 30 and 50. This number is called the target number.

Example: 3, 7, 12, 15, 19

target number  43

The challenge is to use the random numbers and any of the mathematical operations to arrive at the target number. Each random number can be used only once.

Using the previous example, the number 43 could be arrived at in the following ways:

$$(3 \times 7) + 19 + (15 - 12) = 43$$
$$(15 - 12) \times 7 + 19 + 3 = 43$$
$$(15 - 7) \times 3 + 19 = 43$$

If all five random numbers are used to make the target number, the student is awarded 1,000 points; if four numbers are used, 500 points; if three numbers are used, 300 points.

A computer study at the University of California has shown that 86% of the time you can find the target number with any five random numbers.

Students can use calculators to find solutions but should write the combinations and operations they used to arrive at the target number as shown in the examples given above.

NUMBER SENSE

ACTIVITY 2 – Elevator

The game Elevator requires a deck of playing cards with the Jokers removed. The value of the Ace is 1, Jack is 11, Queen is 12 and King is 13. Elevator can be played by groups of three or four students.

To begin play, one of the players shuffles the cards and deals five cards facedown to each player. The remaining cards are placed facedown in a pile. The dealer then turns over two cards from the top of the pile. The numbers turned over are multiplied and the product is the target number.

Each player then looks at his or her cards and using any of the four mathematical operations, tries to make the target number. Each card can be used only once.

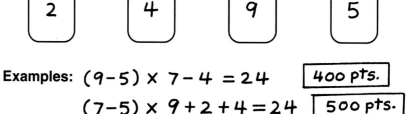

Examples: $(9-5) \times 7 - 4 = 24$ 400 pts.
$(7-5) \times 9 + 2 + 4 = 24$ 500 pts.

Players get 100 points for each number used in the solution.

To find the solutions, students can use calculators or paper and pencil but should try to do mental calculations. After each player finds a solution he or she records the score. The cards are gathered and another player shuffles the cards and deals out the five cards to each player. After four rounds the player with the highest score is the winner.

ACTIVITY 3 – The Date Problem

The Date Problem is a group activity for the entire class. Write today's date on the chalkboard or the overhead. Ask students to write as many equations as they can using these numbers. Tell the students that they must use the numbers from left to right, and that they can use any of the four mathematical operations.

Example: 10 / 29 / 93

NUMBER SENSE

Possible equations:

$$(10+2)-9 = 9+3$$
$$1+(0 \times 2 \times 9) = \sqrt{9} \div 3$$
$$(1 \times 0 \times 2)+9 = \sqrt{9} \times 3$$
$$(1 \times 0)+2+(9 \div 9) = 3$$
$$1+0+2 = (9 \div 9) \times 3$$

MENTAL COMPUTATION

Students like to perform mental computations as quickly and accurately as possible, and there are techniques that help students develop this ability.

Students can enlist the properties of numbers to help them do mental math. The *commutative, associative* and *distributive* properties can help students add or multiply mentally. The *commutative* property of both addition and multiplication states that you can switch the order of the addends or the factors without changing the answer. The *associative* property of addition and multiplication states that you can change the grouping of addends or factors without changing the answer. (6 + 4) + 7 is the same as 6 + (4 + 7). The *distributive* property allows students to break down a problem into simpler parts. 13 x 4 can be written as (10 + 3) x 4 = (10 x 4) + (3 x 4).

OBJECTIVE: Students use the distributive property to compute mentally.

24 x 3 = ?
Think of 24 as (20 + 4)
(20 + 4) x 3 = ?
Some students will "see" this in vertical form:
```
  20 + 4
     x 3
  ──────
  60 + 12
```
$\boxed{60 + 12 = 72}$

ACTIVITY 1
In this activity, students multiply a two-digit number by a one-digit factor using the distributive property.

999 x 6 = ?
Think 999 as (1,000 − 1)
(1,000 − 1) x 6 = ?
6,000 − 6 = 5,994

or

```
  1,000 − 1
        x 6
  ─────────
```
6,000 − 6 = 5,994

ACTIVITY 2
In this activity, students multiply by renaming a number using multiples of ten.

NUMBER SENSE

ACTIVITY 3
In this activity students add by renaming using multiples of ten.

$$15 + 99 = ?$$
Think 99 as $(100 - 1)$
$$15 + (100 - 1) = ?$$
$$(15 + 100) - 1 = ?$$
$$115 - 1 = 114$$

or

$$\begin{array}{r} 100 - 1 \\ + 15 \\ \hline 114 \end{array}$$

OBJECTIVE: Students explore different ways to solve problems using mental computation.

ACTIVITY
Present the students with the problem of solving 96 − 58 in as many ways as possible.

$96 - 58 = ?$	$96 - 58 = ?$	$96 - 58 = ?$
$98 - 60 = 38$	$100 - 62 = 38$	Subtract 50
Add 2 to each number	Add 4 to each number	$96 - 50 = 46$
		Subtract 8
		$46 - 8 = 38$

OBJECTIVE: Students use the commutative and associative properties of numbers for mental computation.

ACTIVITY 1
Demonstrate for your students how to use the commutative (order) and associative (grouping) properties of addition to solve problems with mental computation.

Example:
$$25 + 39 + 75 = ?$$
$$25 + 75 + 39 = ?$$
$$(25 + 75) + 39 = ?$$
$$100 + 39 = 139$$

Using the properties you can change the order of the addends and group them differently.

$$\frac{1}{4} + \frac{5}{6} + \frac{3}{4} = ?$$
$$\left(\frac{1}{4} + \frac{3}{4}\right) + \frac{5}{6} = ?$$
$$1 + \frac{5}{6} = 1\frac{5}{6}$$

$$-7 + 8 + 7 = ?$$
$$(-7 + 7) + 8 = ?$$
$$0 + 8 = 8$$

NUMBER SENSE

ACTIVITY 2

Demonstrate how to use the commutative and associative properties of multiplication to solve problems with mental computation. Look for factors that give products that are easy to use, such as multiples of 10 or 100.

$$25 \times 17 \times 4 = ?$$
$$(25 \times 4) \times 17 = ?$$
$$100 \times 17 = 1,700$$

Mentally multiply 25 x 4, then multiply the product by 17.

$$2 \times 3 \times 7 \times 5 = ?$$
$$(2 \times 5) \times (3 \times 7) = ?$$
$$10 \times 21 = 210$$

Mentally multiply 2 x 5 and 3 x 7, then multiply the products.

COMPUTATIONAL ESTIMATION

As students become involved in many different kinds of estimation activities, they will begin to understand that some answers can be estimates, while other answers require an exact number. An estimation mind-set will help students deal with everyday situations, and will also help them accept estimation as a legitimate part of mathematics. In computational estimation students can use the *rounding strategy* which consists of rounding numbers up or down usually to the nearest 10 or 100.

OBJECTIVE: Students estimate the product of two-digit numbers by rounding to the nearest multiple of ten.

ACTIVITY

Explain to your students that if they round one factor up and round one factor down the estimate is close to the exact product.

```
    58        round to         60
   x 23       round down to   x 20
                              1,200  estimate
```

If both factors are rounded up, the estimate is greater than the exact product.

```
    57        round up to      60
   x 38       round up to     x 40
                              2,400  estimate
```

If students round down both factors, the estimate is less than the exact product.

```
    73        round down to    70
   x 34       round down to   x 30
                              2,100  estimate
```

NUMBER SENSE

OBJECTIVE: Students estimate answers to problems by substituting "nice" numbers.

ACTIVITY

In this activity students substitute numbers to make an estimate of a problem. Use multiples of the divisor to make the estimation easier.

Example: 7)419   substitute 420 for 419

7)420 = 60   estimate

Example: 40)3180   substitute 3,200

40)3200 = 80   estimate

Example: 471
        − 38   substitute 40

         471
        − 40
         431   estimate

Example: 4 × 261   substitute 250 for 261

4 × 250 = 1,000   estimate

NUMBER THEORY

The study of the whole numbers and the various relationships among them is called number theory. Divisibility rules of numbers, prime and composite numbers, exponents and least common multiples are all part of number theory.

OBJECTIVE: Students use the divisibility rules for whole numbers.

ACTIVITY 1

Ask students to explain how they decide if a number is even. Students will argue that the last digit of an even number is 0, 2, 4, 6, or 8. When 2 is a factor of a number, the number is divisible by 2.

Write a four-digit number that is divisible by 2 on the chalkboard. This is an even number. Ask students to give examples of other numbers that are divisible by 2. Then write 235 on the board and ask, "Why isn't 235 divisible by 2?"

NUMBER SENSE

ACTIVITY 2
Have students explain how they identify numbers that are divisible by 5. If the numbers end in 0 or 5, 5 is a factor. If a number is divisible by 5, then 5 is one of its factors. Numbers are divisible by 10 if they end in 0.

Write a four-digit number on the chalkboard that is divisible by 5 but not by 10. Write a four-digit number that is divisible by 10 and 5.

ACTIVITY 3
In this activity, ask students to look for patterns in numbers that have 3 as a factor or are divisible by 3.

Point out to your students that the pattern shows that a number is divisible by 3 only if the sum of the digits is divisible by 3.

Ask your students to write a three-digit number that is divisible by 3.

Then ask: "Why isn't 236 divisible by 3?"
"Is 2,460 divisible by 2? Why?"
"Is 2,460 divisible by 6? Why?"
(If an even number is divisible by 3, it is also divisible by 6.)
"Is 2,512 divisible by 2? by 3?"

ACTIVITY 4
Tell your students that a number is divisible by 9 if the sum of the digits is divisible by 3.

Demonstrate this using the following example:
4,077 is divisible by 9 because the sum of the digits is 18.
$4 + 0 + 7 + 7 = \boxed{18}$
$\boxed{18}$ is divisible by 9

Then ask: "Is 1,251 divisible by 9? Why?"
"Write a four-digit number that is divisible by 9."
"Is 3,456 divisible by 2? by 3? by 9?"

123

NUMBER SENSE

ACTIVITY 5

In this activity, show the students that a number is divisible by 4 if the number formed by the last two digits is divisible by 4.

Put the following examples on the chalkboard:

```
13 [40]   40 ÷ 4 = 10        1,340 IS DIVISIBLE BY 4
28 [92]   92 ÷ 4 = 23        2,892 IS DIVISIBLE BY 4
17 [34]   34 IS NOT          THEREFORE 1,734 IS NOT
          DIVISIBLE BY 4     DIVISIBLE BY 4
```

Ask your students to write a four-digit number that is divisible by 4. A year is a Leap Year if the date is divisible by 4. Is 1996 a Leap Year? Why?

OBJECTIVE: Students identify prime and composite numbers.

ACTIVITY 1

Explain to the students that whole numbers that have no factors other than 1 and the number itself are prime numbers.

Present the following example on paper or on the chalkboard.

WHOLE NUMBER FACTORS

13	8	7	12	5	9
13 × 1	4 × 2	7 × 1	3 × 4	5 × 1	3 × 3
1 × 13	2 × 4	1 × 7	4 × 3	1 × 5	9 × 1
	8 × 1		6 × 2		1 × 9
	1 × 8		2 × 6		
			12 × 1		
			1 × 12		

This demonstrates that 13, 7 and 5 are prime numbers. Their only factors are the number itself and one.

NUMBER SENSE

ACTIVITY 2
Prepare a sheet of numbers 1 to 50 for each student. Tell the students they are to find all the prime numbers less than 50.

SIEVE OF ERATOSTHENES*

* Eratosthenes was a Greek mathematician and astronomer who accurately estimated the circumference of the earth.

Then lead the students through the activity.

- "Cross out 1. One is not prime. It has only one factor."
- "Circle 2, it is prime. Cross out all even numbers."
- "Circle 3, 3 is prime. Cross out all numbers that have 3 as a factor."
- "Circle 5, it is prime. Cross out all numbers that have 5 as a factor."
- "Circle 7, it is prime. Cross out all numbers that have 7 as a factor."
- "Circle 11, it is prime. Cross out all numbers that have 11 as a factor."
- "Circle 13, 17, 19, 23, 29, 31, 37, 41, 43 and 47. All are prime."

The numbers 2, 3, 5, 7, 11, 13, 17, 19, 23, 29, 31, 37, 41, 43 and 47 are prime numbers less than 50.

ACTIVITY 3
Tell your students that whole numbers that have other factors than one and itself are called composite numbers.

Then ask:

- "Which even number is not a composite number?"
- "Why is 51 a composite number?"

Ask your students to write all composite numbers less than 19. Explain that one and zero are neither prime nor composite numbers.

NUMBER SENSE

ACTIVITY 4

Explain that the prime factorization is writing all the prime factors of a number.

Demonstrate the Factor Tree of the number 48.

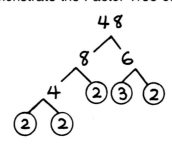

$48 = 2 \times 2 \times 2 \times 3 \times 2$

Explain that you begin with two factors of the number. Continue to factor each one until the last numbers are prime. All of the last circled numbers are the prime factors.

Then show the students that any set of two factors can be used in the first step.

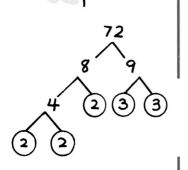

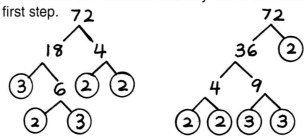

$72 = 2 \times 2 \times 2 \times 3 \times 3$

Now ask the students to write the factorization of 36 using the factor tree.

OBJECTIVE: Students identify exponents and bases and read the short form of equal factors.

ACTIVITY 1

Equal factors can be written in a short form using exponents. The factor is the base. The number of equal factors is the exponent.

For example: $3 \times 3 \times 3 \times 3 \times 3 = 3^5$

$5 \longrightarrow$ exponent
$3 \longrightarrow$ base
$3^5 \longrightarrow$ read "three to the fifth power"

Ask your students to:

Write 6 x 6 x 6 x 6 in exponent form.
Write 2 x 2 x 2 x 2 x 2 x 2 x 2 in exponent form.
Write 4^3 in factor form.
Write 10^4 in factor form.

Then ask: "What is the value of 3^4?"
"Is that the same as 3 x 4?"

NUMBER SENSE

ACTIVITY 2
Students are to look for a pattern in multiples of ten in exponent form.

$10 \times 10 = 10^2$ $10^2 = 100$
$10 \times 10 \times 10 = 10^3$ $10^3 = 1{,}000$
$10 \times 10 \times 10 \times 10 = 10^4$ $10^4 = 10{,}000$

Ask the students: "What pattern exists between powers of ten and the zeroes in the products?"

"Now, write 1,000,000 in exponent form."

OBJECTIVE: Students find the least common multiple of several numbers.

ACTIVITY 1
Draw a Venn Diagram on the chalkboard, as shown to illustrate common multiples. Ask the students to give multiples of 6 and 8. For example, 12, 16, 30, 40, 24, 18, 56, 48.

Write the numbers in the Venn Diagram as students tell where they belong. Is it apparent that 24 and 48 are multiples of both 6 and 8?

Then ask the students:
"What is the next largest common multiple of 6 and 8?" (72)
"What is the least common multiple of 6 and 8?" (24)

ACTIVITY 2
Demonstrate how to use Euclid's method to find the least common multiple. Begin by dividing both numbers by two if possible or by the next prime number. Continue dividing by primes that are common factors of the number.

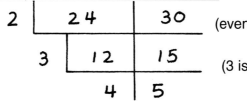

Least common multiple of 24 and 30 is 2 x 3 x 4 x 5, or 120.

Find the least common multiple of 35 and 28.

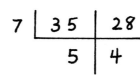

2, 3, 5 cannot be used as divisors. Begin with the common factor 7. There are no more common factors. So, 7 x 5 x 4 = 140 – the least common multiple.

127

NUMBER SENSE

OBJECTIVE: Students use the rules for the order of operations.

ACTIVITY 1

Put the following two examples on the chalkboard to illustrate why the rules for the order of operations are needed.

Incorrect   Correct

Present the rule that if a problem contains a series of operations, first perform the multiplication and division in the order in which they occur. Then do the addition and subtraction in order from left to right.

Example:

$$81 \div 9 - 3 \times 2 + 4 =$$
$$9 - 6 + 4 =$$
$$3 + 4 = 7$$

Ask your students to perform the indicated operations following the order of operation rules. Hint: Sometimes it helps to put parentheses around the pairs to be done first.

$$45 \div 5 + 20 \times 3$$
$$3 + 72 \div 8 \times 3 - 6 \times 2$$
$$24 \div 6 \times 3 - 8 \div 4 + 6$$

ACTIVITY 2

Explain to the students that when parentheses occur, do that computation first. Then follow the rule of multiplication and division in order from left to right. Then add or subtract in order from left to right.

Use the following examples:

$$8 - 4 \times (5-2) \div 3 - 2$$
$$3 + (12 \div 4) \times 3 + (6 \times 8) \div 2 - 5$$
$$16 \div 4 \times 2 + 5(30 \div 6) - 15 \div 3$$